Table of Contents

1 Introduction

The Land Use, Land-Use Change, and Forestry (LULUCF) Guidance for GHG Project Accounting (LULUCF Guidance) was developed by the World Resources Institute to supplement the GHG Protocol for Project Accounting (Project Protocol). This document provides more specific guidance and uses more appropriate terminology and concepts to quantify and report GHG reductions from LULUCF project activities. The LULUCF Guidance was written in consultation with and reviewed by many stakeholders, similar to the process used to develop the Project Protocol.

The LULUCF Guidance is intended to be used in conjunction with—not in place of—the Project Protocol, so project developers should read the Project Protocol first in order to become familiar with the general framework for GHG project accounting, as most of this information is not repeated in the LULUCF Guidance.

1.1 About the Greenhouse Gas Protocol Initiative

The Greenhouse Gas Protocol Initiative is a partnership of businesses, nongovernmental organizations (NGOs), governments, academics, and others convened by the World Business Council for Sustainable Development (WBCSD) and the World Resources Institute (WRI). Launched in 1998, the initiative's mission is to create internationally accepted greenhouse gas (GHG) accounting and reporting standards and/or protocols and to promote their broad adoption.

The GHG Protocol Initiative contains two separate but linked modules:
- The GHG Protocol Corporate Accounting and Reporting Standard (Corporate Standard), revised edition, published in March 2004.

- The GHG Protocol for Project Accounting (Project Protocol), published in November 2005.

1.2 About the GHG Protocol for Project Accounting

The Project Protocol provides specific principles, concepts, and methods for quantifying and reporting GHG reductions—that is, the decreases in GHG emissions or the increases in GHG removals—from climate change mitigation projects (GHG projects). The Project Protocol is the culmination of four years of dialogue and consultation with many stakeholders and uses the knowledge and experience of a wide range of experts.

The Project Protocol
- Provides a credible and transparent approach to quantifying and reporting GHG reductions from GHG projects.

- Enhances the credibility of GHG project accounting by means of common accounting concepts, procedures, and principles.

- Provides a platform for harmonizing the different project-based GHG initiatives and programs.

The Project Protocol clarifies the requirements for quantifying and reporting GHG reductions and offers guidance and principles for meeting those requirements. Although the requirements are extensive, there is considerable flexibility in the ways of meeting them, because GHG project accounting means making decisions directly related to GHG programs' policy choices, namely, trade-offs among environmental integrity, program participation, program development costs, and administrative burdens. Furthermore, because the Project Protocol is not intended to promote any specific programs or policies, accounting decisions related to program or policy design are left to the discretion of its users.

1.3 About the LULUCF Guidance

The LULUCF Guidance is designed to facilitate the use of the Project Protocol for LULUCF project activities. Its format is similar to that of part II of the Project Protocol, but it contains specific LULUCF guidance. Although the LULUCF Guidance may be used for all LULUCF project activities, it focuses on two project types: reforestation[1] and forest management. This guide also can be used for avoided deforestation project activities, although they are not explicitly discussed.

The LULUCF Guidance has no requirements; it simply describes and illustrates, using one example[2]: how the requirements in the Project Protocol for reforestation and forest management project activities should be fulfilled. It highlights those elements for which LULUCF project activities may need approaches slightly different from those in the Project Protocol. In addition, this document points out areas where GHG programs may improve the practicality of these methodologies, by reducing the uncertainty and transaction costs of developing GHG projects while at the same time enhancing the projects' environmental integrity.

1.4 Who Should Use the Project Protocol and the LULUCF Guidance?

The Project Protocol and the LULUCF Guidance are written for project developers, but administrators or designers of initiatives, systems, and programs that incorporate GHG projects, as well as third-party verifiers for such programs and projects, may find it useful as well. Indeed, anyone wanting to quantify the GHG reductions resulting from GHG projects may use the Project Protocol, and those considering reforestation or forest management projects should also read the LULUCF Guidance. GHG projects are undertaken for a variety of reasons, such as generating officially recognized GHG reduction "credits" to meet mandatory emission targets, obtaining recognition for voluntary GHG reductions, and offsetting GHG emissions to meet internal company targets for public recognition or other internal strategies. Although the Project Protocol and the LULUCF Guidance are intended for all these purposes, their use does not guarantee a particular result regarding quantified GHG reductions or the acceptance or recognition by GHG programs that have not adopted their provisions. Therefore, users are encouraged to consult with relevant

FIGURE 1.1 Steps in Accounting and Reporting GHG Reductions from LULUCF Projects

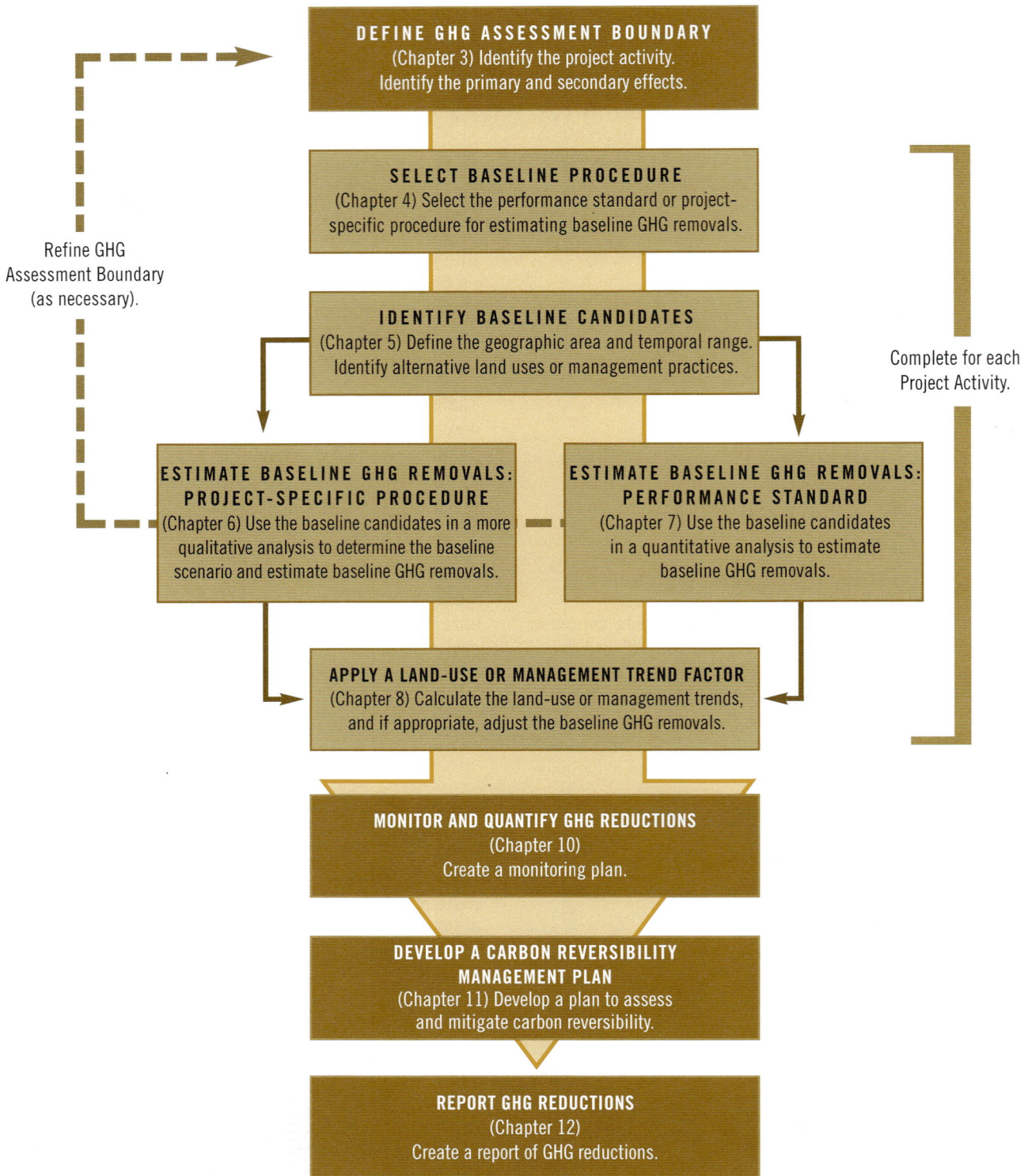

DEFINE GHG ASSESSMENT BOUNDARY
(Chapter 3) Identify the project activity.
Identify the primary and secondary effects.

SELECT BASELINE PROCEDURE
(Chapter 4) Select the performance standard or project-specific procedure for estimating baseline GHG removals.

IDENTIFY BASELINE CANDIDATES
(Chapter 5) Define the geographic area and temporal range.
Identify alternative land uses or management practices.

Refine GHG
Assessment Boundary
(as necessary).

Complete for each
Project Activity.

ESTIMATE BASELINE GHG REMOVALS: PROJECT-SPECIFIC PROCEDURE
(Chapter 6) Use the baseline candidates in a more qualitative analysis to determine the baseline scenario and estimate baseline GHG removals.

ESTIMATE BASELINE GHG REMOVALS: PERFORMANCE STANDARD
(Chapter 7) Use the baseline candidates in a quantitative analysis to estimate baseline GHG removals.

APPLY A LAND-USE OR MANAGEMENT TREND FACTOR
(Chapter 8) Calculate the land-use or management trends, and if appropriate, adjust the baseline GHG removals.

MONITOR AND QUANTIFY GHG REDUCTIONS
(Chapter 10)
Create a monitoring plan.

DEVELOP A CARBON REVERSIBILITY MANAGEMENT PLAN
(Chapter 11) Develop a plan to assess and mitigate carbon reversibility.

REPORT GHG REDUCTIONS
(Chapter 12)
Create a report of GHG reductions.

Chapter 9: Estimating and Quantifying Carbon Stocks is not a step, but is useful for chapters 6 through 10.

programs or other interested parties regarding policy-relevant accounting decisions. Users without guidance on these decisions should strive for maximum transparency when justifying such decisions and fulfilling the Project Protocol's requirements.

1.5   Overview of the LULUCF Guidance

The LULUCF Guidance contains four parts. Part I provides background information; part II is based on the Project Protocol and offers specific guidance for LULUCF project activities; part III uses an example to illustrate the guidance outlined in part II; and part IV contains the annexes.

PART I: CONCEPTS AND PRINCIPLES

• **Chapter 1: Introduction.** This chapter introduces the GHG Protocol Initiative, the Project Protocol, and the LULUCF Guidance; outlines the uses and limitations of the Project Protocol and the LULUCF Guidance; and provides an overview of the LULUCF Guidance and specific LULUCF issues not covered in the LULUCF Guidance.

• **Chapter 2: Key LULUCF Accounting Concepts and Principles.** This chapter defines the key accounting concepts specific to LULUCF GHG project accounting and outlines the principles found in the Project Protocol for project accounting.

PART II: GHG REDUCTION ACCOUNTING AND REPORTING

Figure 1.1 shows the steps in calculating and reporting GHG reductions from LULUCF projects and where they are addressed in this guidance.

• **Chapter 3: Defining the GHG Assessment Boundary.** This chapter explains how to identify the GHG sources and sinks that should be considered when quantifying GHG reductions. First, the GHG project should be separated into one or more "project activities." In addition to its primary effects—the specific changes in GHG removals that a project activity is designed to achieve—a project's activities may result in unintended changes in GHG emissions elsewhere, that is, secondary effects. By defining the GHG assessment boundary (which includes all primary and significant secondary effects associated with each project activity), project developers can identify the sources and sinks to be considered when calculating the GHG reductions.

5

The two procedures for estimating "baseline GHG removals"—that is, the GHG removals that are compared with the project activity GHG removals to determine the GHG reductions—are (1) the performance standard procedure and (2) the project-specific procedure.

- **Chapter 4: Selecting a Baseline Procedure.**
This chapter describes each procedure and explains which one should be used for particular forestry projects and conditions. The choice of procedure affects the identification of baseline candidates (chapter 5) and the estimation of baseline GHG removals (chapters 6 and 7).

- **Chapter 5: Identifying the Baseline Candidates.**
This chapter shows how to identify baseline candidates; that is, the alternative land uses or management practices to consider and analyze when estimating baseline GHG removals. This is done by defining a relevant area and time frame from which the baseline candidates are identified. It is important that this be done accurately, as both baseline procedures (chapters 6 and 7) use the baseline candidates to derive the baseline GHG removals. Baseline candidates are also important if a land-use or management trend is used to adjust the baseline GHG removals (chapter 8).

- **Chapter 6: Estimating the Baseline GHG Removals— Project-Specific Procedure.**
This chapter explains how to estimate baseline GHG removals using the "project-specific" procedure. The first part of this procedure identifies a "baseline scenario" from the list of baseline candidates in chapter 5, and the second part estimates the baseline GHG removals associated with the baseline scenario. These baseline GHG removals are then compared with the project activity GHG removals to calculate the total GHG reduction, explained in chapter 10.

- **Chapter 7: Estimating the Baseline GHG Removals— Performance Standard Procedure.**
This chapter shows how to estimate baseline GHG removals using the "performance standard" procedure. To do this, the baseline GHG removals are estimated by ranking the baseline candidates based on their GHG removals and then selecting the better-than-average GHG removals (the *performance standard*) to use as baseline GHG removals. As with the project-specific procedure, the baseline GHG removals are compared with the project activity GHG removals to calculate the total GHG reduction, explained in chapter 10.

- **Chapter 8: Applying a Land-Use or Management Trend Factor.**
This chapter gives instructions for estimating the rate at which land-use or management changes are occurring in a region, that is, the land-use or management trend factor. This factor can be applied to both the performance standard and the project-specific baseline procedures to derive a more accurate estimate of baseline GHG removals over time, thereby ensuring that the baseline GHG removals more closely reflect the region's changing conditions.

- **Chapter 9: Estimating and Quantifying Carbon Stocks.**
This chapter describes the components necessary to estimate or quantify carbon stocks and lists various resources to use.

- **Chapter 10: Monitoring and Quantifying GHG Reductions.**
This chapter shows how to monitor and quantify GHG reductions from LULUCF project activities. It also supplies the equations to calculate the total GHG reductions using baseline GHG removals (estimated in chapters 6 and 7) and project activity GHG removals.

- **Chapter 11: Carbon Reversibility Management Plan.**
This chapter explains how to develop a management plan for preventing and mitigating the effects of intentional or unintentional carbon reversals (e.g., harvesting activities, forest fires, insect infestation). Intentional reversals should be factored into the ex-ante assessment of the project's GHG reduction in chapter 10, taking into account any compensating mechanisms incorporated into the project design. The project reduction is then calculated, based on the monitoring of actual carbon stored by the project. The carbon reversibility management plan should make sure that the project reductions are actually achieved and are not intentionally or unintentionally reversed.

- **Chapter 12: Reporting GHG Reductions and Net Carbon Stocks.**
This chapter describes how to report transparently the GHG reductions calculated in chapter 10 and to apply to LULUCF projects the reporting requirements listed in the Project Protocol.

PART III: EXAMPLE
Part III offers an example that illustrates the application of this guidance: Nipawin Afforestation Project. Other examples are available on the GHG Protocol web site (www.ghgprotocol.org).

PART IV: SUPPLEMENTARY INFORMATION
There are four annexes in this document, providing project developers with specific information about life-cycle assessments and secondary effects, GHG programs' definitions of forest, afforestation, etc., and more detailed QA/QC procedures.

1.6   Issues Not Addressed in the LULUCF Guidance

Like the Project Protocol, the LULUCF Guidance does not address sustainable development, stakeholder consultation, ownership of GHG reductions, confidentiality, and verification, although it does address uncertainty, albeit briefly. In addition, the guidance does not explicitly discuss the following three issues specific to LULUCF projects:

- **Crediting:** Crediting is the provision of credits based on the GHG removals over a given period of time. Although how crediting should be handled for LULUCF project activities has been widely discussed in the literature, this is a policy decision, and therefore the LULUCF Guidance does not address it.

- **Ecological considerations and co-benefits from LULUCF project activities:** Land-use management and ecological conditions (e.g., water and biodiversity considerations, community livelihoods) are important components of LULUCF project activities and should be considered when developing a project. But they are beyond the scope of this document and so are not covered.

- **Wood products:** Various expert groups are still debating accounting conventions for the carbon stored in wood products. In the 1996 Good Practice Guidance for developing national inventories, the Intergovernmental Panel on Climate Change (IPCC) did not include the carbon stored in wood products after harvesting. For GHG accounting, this means that all carbon stored in the trees was considered emitted back into the atmosphere during harvest and that any carbon that may have remained in wood products was thus not accounted for. This methodology however is changing. The 2006 IPCC guidelines include a variety of methods to address carbon stored in wood products, and GHG programs like the California Climate Action Registry and the U.S. Department of Energy's Voluntary Reporting of Greenhouse Gases (1605[b]) Program provide guidance on accounting for wood products. Other organizations such as the National Council for Air and Stream Improvement (NCASI), working with the International Council of Forest Paper Associations (ICFPA), have a manual entitled *ICFPA/NCASI Tools for Calculating Biomass C Stored in Forest Products In-Use — Instruction Manual v1.0.* Other guidance documents may also exist. In the short term, however, if and how the wood products pool is incorporated into GHG accounting for LULUCF project activities will most likely continue to be policy decisions made by individual GHG programs and so are not addressed here.

7

NOTES
[1] This document uses the term reforestation broadly and generically to cover any activities that include planting trees in nonforested areas. For more information about program-specific definitions of afforestation, reforestation, and how these terms are used here, see annex B.

[2] Others are available at www.ghgprotocol.org.

2 Key LULUCF Accounting Concepts and Principles

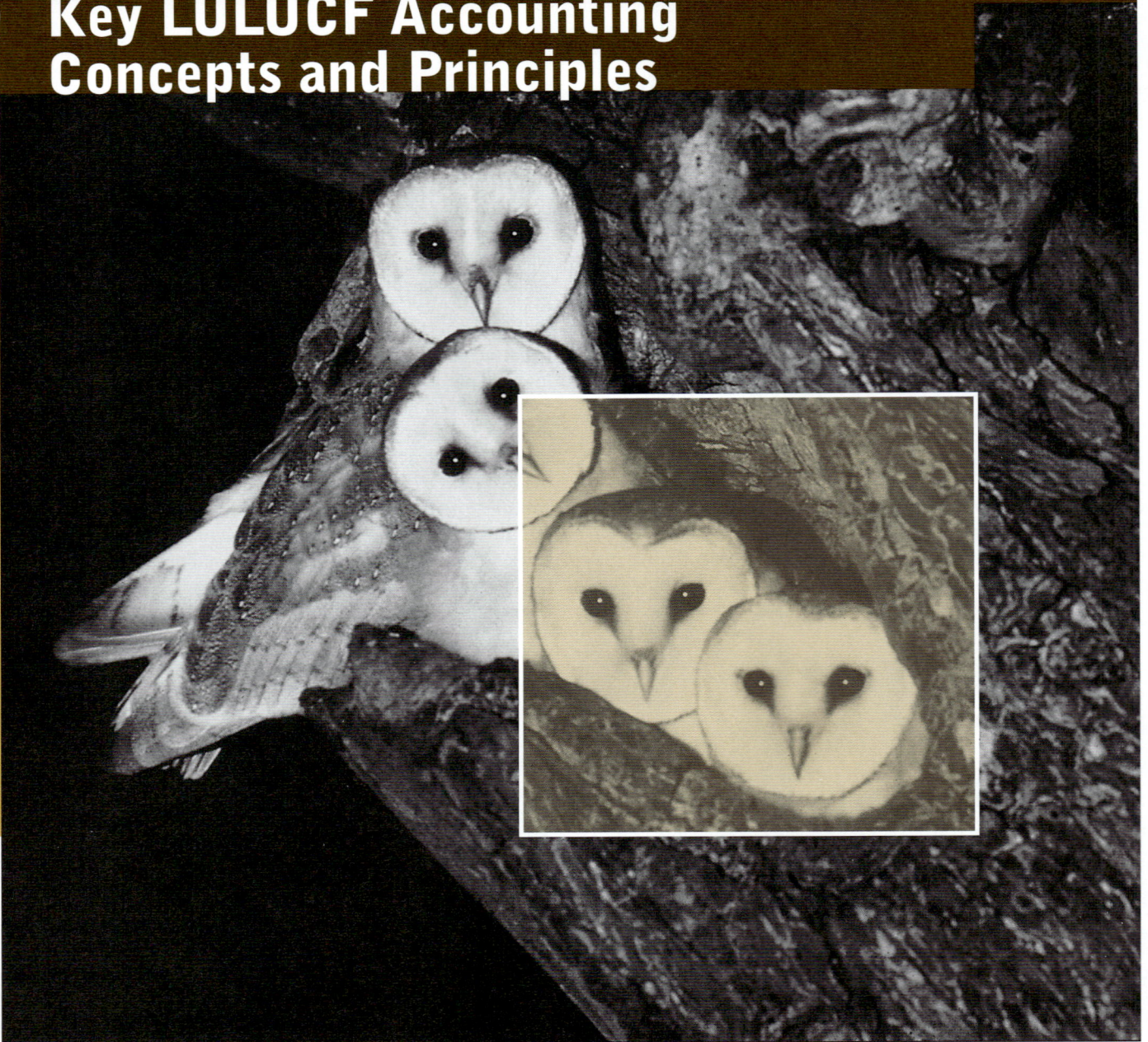

A number of key concepts must be understood to account for GHG reductions from LULUCF projects. This chapter explains the importance of the concepts and describes how and where they are used in the LULUCF Guidance. The concepts presented here are also defined in the glossary. In addition, this chapter outlines the principles presented in the Project Protocol.

2.1 Key Concepts

Several accounting terms and definitions pertain specifically to LULUCF project activities or have been adapted from the Project Protocol for LULUCF project activities.[1] This chapter defines and describes the following terms according to their importance to LULUCF carbon accounting:

- Carbon stocks, change in carbon stocks and GHG removals

- GHG effects

- Baseline candidates and baseline procedures

- Land-use or management trend factor

- Variability

- Uncertainty

- Permanence

- Additionality

2.1.1 CARBON STOCKS, CHANGE IN CARBON STOCKS, AND GHG REMOVALS

To quantify GHG reductions and capture the data required to transparently report the achievements of a LULUCF project, project developers need:

1. The change in carbon stocks[2] between two time periods for the baseline and the project activity per unit of land area (t C/ha). The change in carbon stocks may be either negative or positive, depending on the activities taking place, for example, growing trees versus harvesting trees.

The change in carbon stocks is translated into GHG removals by multiplying t C/ha by $\frac{44}{12}$ (the ratio of the molecular weight of carbon dioxide [CO_2] to the molecular weight of carbon) to get t CO_2/ha, which also allows the project developer to easily include nonbiological GHG emissions into the calculation of the GHG removals.

2. The total baseline carbon stocks and total project activity stocks in t C. This is used to compare the total carbon stored, as well as to track the longevity of the carbon stored (permanence).

The GHG reduction is then calculated by finding the difference between the project GHG removals and the baseline GHG removals in the same time period (see figure 2.1).

The change in carbon stocks (sometimes referred to as flux) usually reflects the growth rate of trees/vegetation and the dynamics of carbon in the soil and other pools between two time periods, as well as the corresponding carbon that is stored. If the change in carbon stocks is negative, then the forest is sometimes called a source, since it is emitting or losing carbon. If the change in carbon stocks is positive during a given time period, the forest is often referred to as a sink, since it is absorbing carbon.

This guidance uses the terms *change in carbon stocks* and *GHG removals*. The former is found in terms of carbon (e.g., t C/ha) and is only used when looking at biological

9

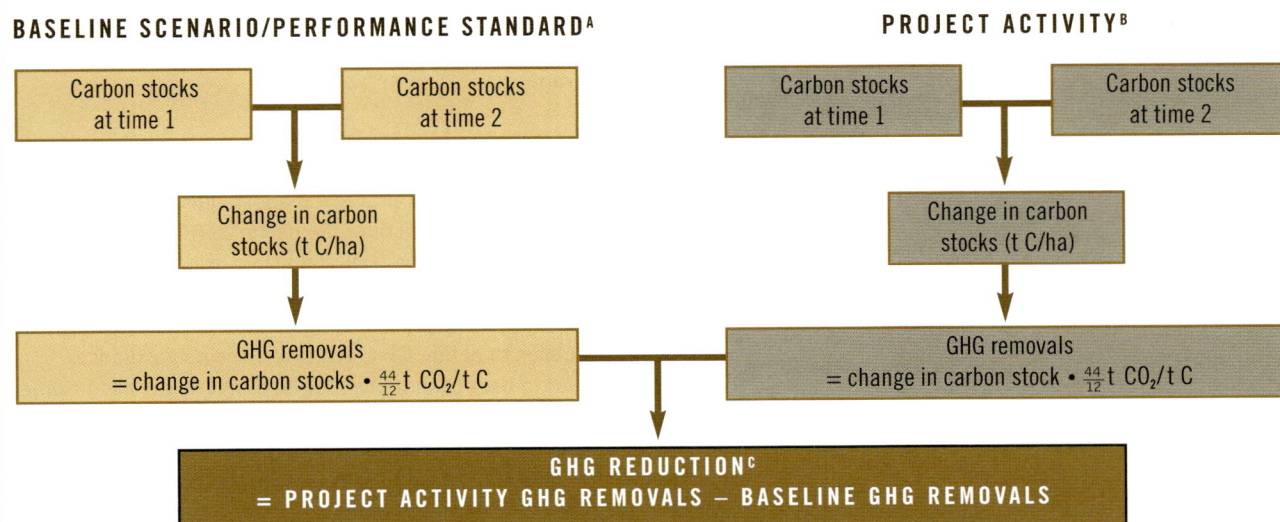

FIGURE 2.1 Calculating GHG Removals Using Carbon Stocks

BASELINE SCENARIO/PERFORMANCE STANDARD[A]

PROJECT ACTIVITY[B]

Carbon stocks at time 1

Carbon stocks at time 2

Change in carbon stocks (t C/ha)

GHG removals = change in carbon stocks $\cdot \frac{44}{12}$ t CO_2/t C

Carbon stocks at time 1

Carbon stocks at time 2

Change in carbon stocks (t C/ha)

GHG removals = change in carbon stock $\cdot \frac{44}{12}$ t CO_2/t C

GHG REDUCTION[C]
= PROJECT ACTIVITY GHG REMOVALS − BASELINE GHG REMOVALS

[A] See section 2.1.3 for an explanation of the baseline scenario versus performance standard.
[B] Only the primary effects are considered here.
[C] GHG reductions are calculated in CO_2eq.

effects, and the latter is found in terms of carbon dioxide (e.g., t CO_2/ha).

Calculating the GHG removals is important for crediting purposes because credit is given for the difference between project activity GHG removals and baseline GHG removals and not the total amount of carbon stored at any given time. Total carbon stocks however can be used to compare forests with less biomass (e.g., younger or less dense plantings) versus forests with more biomass (e.g., older, mature forests). Carbon stocks also enable stakeholders to more easily assess the storage of carbon in biomass and soils over time (permanence). Therefore this document recommends reporting both metrics.

2.1.2 GHG EFFECTS

GHG effects are the GHG removals and emissions caused by a project activity and are either primary effects or secondary effects.

Primary Effects

The *primary effect* for LULUCF project activities is all the biological carbon stock changes caused by the project activity on the project site. It is defined as the difference between project activity GHG removals and baseline GHG removals, the latter which are determined using either of the baseline procedures described in chapters 6 and 7. Although a given unit area of land may act as a source or a sink at any given time throughout the GHG project's life, ultimately the project activity is expected to act as a net sink.

Secondary Effects

The *secondary effects* for LUCUCF project activities are the changes in nonbiological GHG emissions caused by the project activity and any biological carbon stock changes that occur off the project site, for example, from a market responses. Secondary effects are unintended in terms of the GHG reduction, although they may be an integral part of the project activity, for example, the GHG emissions related to site preparation for planting or from harvesting wood fiber.

Secondary effects are typically small compared with the project activity's primary effect. In some cases, however, they may cause the project site to act as a source of GHG emissions during certain time periods, or they may even invalidate the entire value of the primary effect over time. The two categories of secondary effects are:
- *One-time effects:* Changes in GHG emissions associated with the implementation or termination of a project activity.

- *Upstream and downstream effects:* Recurring changes in GHG emissions associated with inputs to the project activity (upstream), such as GHG emissions from any vehicles used in the maintenance of the project site, or the products of the project activity (downstream), such as mobile combustion emissions from transporting harvest fiber to the mill.

Some upstream and downstream effects may cause market responses to changes in the supply and/or demand for the project activity's inputs or products. This is often referred to as leakage. For instance, a market response would be an increase in fiber production in a new location when production at the project site falls or ceases. For more information about market responses and other secondary effects, see chapter 3.

2.1.3 LULUCF BASELINE CANDIDATES AND THE BASELINE PROCEDURES

The GHG reductions associated with a LULUCF project are quantified according to a reference level of GHG removals. That reference level is calculated using *baseline candidates,* the alternative land uses or management practices (and their associated GHG removal levels) that could be implemented on the project activity site. Baseline candidates are identified by exploring potential land uses or management practices within a specified geographic area and over a defined temporal range. Once feasible alternatives have been identified, one of two different procedures may be used to derive baseline GHG removals from the baseline candidates. The procedures, each using the term *baseline candidates* slightly differently, are introduced next. Figure 2.2 illustrates how baseline candidates differ between the two baseline procedures.

Project-Specific Procedure

The *project-specific procedure* compares the land-use or management practice alternatives—the baseline candidates—to identify the baseline candidate that represents the baseline scenario (see chapter 6). The GHG removals associated with that baseline scenario become the reference removal level and are compared with the project activity GHG removals to calculate the LULUCF GHG reduction. When using this procedure, the term *baseline candidate* refers to the types of feasible land use or management practice identified in a specified area, for example, cropland or pasture.

Performance Standard Procedure

Rather than identifying a single baseline scenario, i.e., land-use or management practice against which to measure the LULUCF GHG reduction, the performance standard procedure derives baseline GHG removals using a numerical analysis of the GHG removals from all baseline candidates. Instead of using types of feasible land uses or management practices to define GHG removals, the performance standard uses each individual land unit in the geographic area (e.g., each hectare) and corresponding GHG removals as the baseline candidates. These baseline candidates are then ranked according to their GHG removals, from lowest to highest. Using a predefined better-than-average stringency level (see chapter 7), the baseline GHG removals are derived from this spectrum of potential GHG removals.

FIGURE 2.2 Relationship between Baseline Candidates and Baseline Procedures

The following is a diagram of a geographic area whose identified land uses have remained the same over the last twenty-five years:

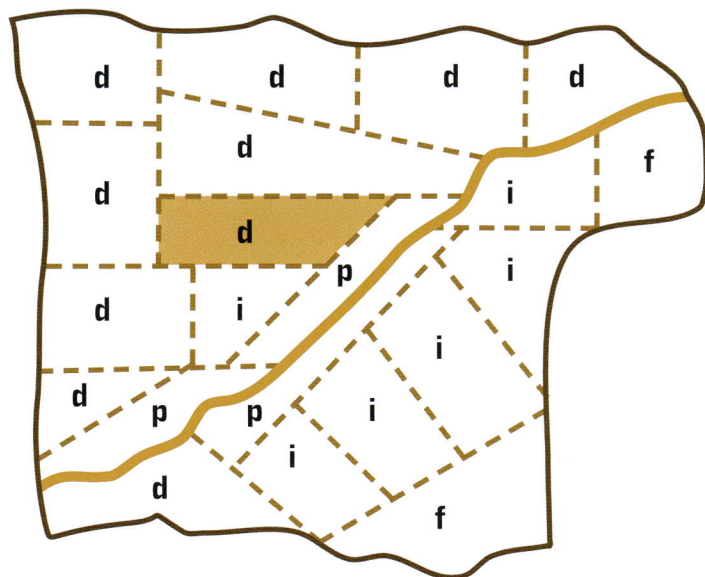

Types and Quantity of Land Use in Area

TYPES	QUANTITY
p = pasture	500 hectares
f = plantation forestry	523 hectares
d = dryland cereal crops	5050 hectares
i = irrigated cereal crops	2036 hectares

Geographic area ——————

River ——————

The shaded area is the project site.

After looking at the biophysical conditions of the project site and the regulatory factors in the area, the following alternative land uses were identified in the geographic area:
- Pasture
- Plantation forestry
- Dryland cereal crops
- Irrigated cereal crops

In the **project-specific procedure**, the baseline candidates are:
- Baseline candidate 1: Pasture
- Baseline candidate 2: Plantation forestry
- Baseline candidate 3: Dryland cereal crops
- Baseline candidate 4: Irrigated cereal crops

Using the **performance standard procedure**, each hectare of land in the geographic area represents a baseline candidate. In this case, the baseline candidates are:
- Baseline candidates 1 to 500: Pasture
- Baseline candidates 501 to 1023: Plantation forestry
- Baseline candidates 1024 to 6073: Dryland cereal crops
- Baseline candidates 6074 to 8109: Irrigated cereal crops

11

2.1.4 LAND-USE OR MANAGEMENT TREND FACTORS

A *land-use or management trend factor* (abbreviated in this document to land-use trend factor) estimates the rate at which land-use or management changes are occurring within the geographic area during the temporal range (both identified in chapter 5). The land-use trend factor is applied to the baseline GHG removals to adjust the GHG removals to reflect the changing land-use or management practices in an area.

Whether it is appropriate to apply a land-use trend factor depends on the nature of the baseline candidates and the amount of rigorous data available to develop a relevant factor. If the baseline candidates represent discrete land uses or types of management practices, a land-use trend

factor should be considered. But if the project activity represents a change in the level or intensity of a management practice or land use, then the land-use trend factor is most likely not applicable.

Reforestation Project Activities

Typical baseline candidates in reforestation projects are discrete land uses, for example, forest, cropland, or pasture. Therefore, a land-use trend factor could be used for all reforestation project activities.

Forest Management Project Activities

The project activity for many forest management projects represents a change in the level or intensity of a management practice, for example, lengthening the harvest rotation or changing the degree of forest thinning. In these situa-

tions, identifying discrete practices across the landscape may not be possible, and therefore estimating a land-use trend factor may not be appropriate.

If the project activity represents a change in the type of management practice providing a given service or product, then the land-use trend factor should be considered. Mechanical weeding, chemical weeding, and thinning forest management practices are examples of discrete forest management practices to enhance the growth of desirable wood fiber species in a forest. It may be possible to determine the rate at which landowners are moving from previous practices to practices increasing the GHG removals outside a specific climate change or other incentive program. For more information about when and how to apply the land-use trend factor, see chapter 8.

2.1.5 VARIABILITY
Carbon stock accounting uses two types of "variability":

1. The variability of carbon stocks and/or changes in carbon stocks *within* a type of baseline candidate or group of similar baseline candidates. For example, one category of baseline candidate encompasses 1000 hectares, and the soil carbon has been measured fifty times over this area. The soil carbon measurements range between 20 and 100 t C/ha in a given year and represents the variability within a given type of baseline candidate. The same could occur when measuring the change in carbon stocks.

2. The variability of carbon stocks and change in carbon stocks *across* different baseline candidates. For example, one type of baseline candidate has low carbon stocks (e.g., 50 t C/ha in a given year), and another type of baseline candidate has higher carbon stocks (e.g., 300 t C/ha in the same year).

The question of variability becomes important at several stages in the accounting process. Many of the criteria used to define the geographic area should help minimize the variability *within* a type of baseline candidate, for example, climatic and soil conditions. Characterizing the distinctions among similar types of baseline candidates may also help reduce the variability, for example, distinguishing land uses using different tillage practices as different types of baseline candidates, or distinguishing forest tracks with different previous forest disturbances as different types of baseline candidates. Determining the degree of variability within baseline candidates is important when quantifying the baseline carbon stocks and changes in carbon stocks and should be considered when addressing issues of accuracy, precision, and conservatism. High variability within baseline candidates may result in high uncertainty levels in the quantified baseline carbon stocks or carbon stock changes. For more information, see chapter 9.

The variability of the change in carbon stocks across the different types of baseline candidates is important to the performance standard when deciding which stringency level to apply. When baseline candidates have very similar carbon stock changes, it may be difficult to select a stringency level that differentiates various land use or management practices. The Nipawin Afforestation example illustrates this situation.

2.1.6 UNCERTAINTY
Uncertainty in the estimation or quantification of the GHG reductions can lower their perceived value to buyers or other stakeholders. Uncertainty is encountered at many stages in the accounting process, from the measurement or estimation of carbon in various carbon pools and the change in carbon stocks expected in each pool, to the longevity of the carbon stock storage (permanence). Clearly indicating where the uncertainty exists and how it has been addressed—for example, by making conservative estimates or creating a management plan to minimize unintentional releases of carbon dioxide to the atmosphere—helps assure buyers, investors, and other stakeholders of the LULUCF project's integrity. Uncertainty, however, is not limited to LULUCF projects.

2.1.7 PERMANENCE

Permanence refers to the longevity of a carbon pool and the stability of its stocks over time. How a particular GHG program addresses permanence is a policy decision and therefore is not covered in detail in this document. Chapter 11, however, offers a carbon reversibility management plan showing project developers how to document the risks to the carbon stored in the different pools and how they can mitigate and monitor those risks. The LULUCF Guidance also describes how to track both the total carbon stocks and the GHG removals over time, thereby allowing stakeholders to see how the net storage of carbon on the land is progressing.

2.1.8 ADDITIONALITY

Project-based GHG reductions are quantified relative to baseline GHG removals, which are derived either from an identified baseline scenario or by using a performance standard that serves the same function as a baseline scenario. Although a project activity is generally assumed to differ from its baseline scenario, a project activity (or the same land-use or forest management practice it employs) sometimes may have been implemented "anyway." In these cases, the project activity and its baseline scenario are effectively identical. Such a project activity may appear to increase GHG removals relative to historical removal rates. But compared with its baseline scenario, the project activity does not generate GHG reductions. GHG programs should count only GHG reductions from project activities that differ from—or are additional to—their baseline scenarios. Differentiating a project activity from its baseline scenario is often referred to as *determining additionality*. Even though the basic concept of additionality may be easy to understand, there is no common agreement on how to prove that a project activity and its baseline scenario are different. The two baseline procedures (project-specific and performance standard) reflect two different methodological approaches to additionality.

2.2 Principles

The following principles were taken from, and are described more thoroughly in, the Project Protocol. They underpin all aspects of the accounting, quantification, and reporting of project-based GHG reductions. Their purpose is to guide decisions when the Project Protocol and the LULUCF Guidance afford flexibility or discretion or when the requirements and/or guidance are ambiguous with respect to a particular situation. The application of these principles will help ensure the credibility and consistency of efforts to quantify and report project-based GHG reductions according to the Project Protocol and the LULUCF Guidance.

The principles are derived in part from accepted financial accounting and reporting principles.

Relevance: Use data, methods, criteria, and assumptions appropriate to the intended use of the reported information.

Completeness: Consider all relevant information that may affect the accounting and quantification of GHG reductions, and complete all requirements.

Consistency: Use data, methods, criteria, and assumptions that allow meaningful and valid comparisons.

Transparency: Provide clear and sufficient information for reviewers to assess the credibility and reliability of GHG reduction claims.

Accuracy: Reduce uncertainties as much as is practical.

Conservativeness: Use conservative assumptions, values, and procedures when uncertainty is high, and do not overestimate GHG reductions.

NOTES

[1] This discussion is not meant to replace chapter 2, Key GHG Project Accounting Concepts, of the Project Protocol, which contains a complete description of the concepts used in this document.

[2] Carbon stock refers to all carbon stored in the measured carbon pools.

3 Defining the GHG Assessment Boundary

The GHG assessment boundary encompasses all GHG sources or sinks associated with all primary effects and any significant secondary effects arising from each project activity in the GHG project. Defining the GHG assessment boundary allows the project developer to decide which carbon pools and other potential sources or sinks to include in the quantification of the GHG reductions.

When defining a GHG assessment boundary,

- Identify the GHG project activity (or activities).

- Identify the primary and secondary effects associated with each project activity.

- Analyze those secondary effects to determine which are relevant to estimating and quantifying the GHG reductions.

This chapter closely follows chapter 5 of the Project Protocol, elaborating on each section as it applies to LULUCF project activities in general and to reforestation and forest management project activities in particular.

3.1 Identifying the Project Activities

The Project Site

The project site is the physical area where the project activity will take place. The first step in identifying the GHG assessment boundary for a LULUCF project is ensuring that the location and size of the project site are accurately and completely defined. The location of the project site is important because it is the basis for a project activity's carbon stocks and carbon stock changes. Unlike other project types, the carbon stocks for the same type of project activity are different at different locations because of variations in a location's characteristics. The potential storage of carbon also differs depending on the size of the project site.

The Project Activity

Once the location and size of the project site have been clearly described, the second step is identifying the project activity or activities associated with the GHG project. The project activity is the specific action or intervention targeted at changing GHG emissions or removals.

When defining the project activity(ies), the project developer should consider how the GHG project will differ from current land-use trends for similar types of land. For example, a project developer may be trying to increase the carbon storage on a forested site through better forest management practices. But if the current land-use trend in the area is selling forested lands for development or agricultural use, the project activity may instead be forest conservation (i.e., to avoid deforestation) rather than forest management.

Reforestation Project Activities

The project activity for reforestation projects is changing the land use and land cover from a nonforest use to a forest use in order to enhance carbon storage. Reforestation project activities involve planting or restoring trees on lands that are not considered forest,[1] such as agricultural or abandoned land.

Forest Management Project Activities

Forest management project activities may increase overall GHG removals and reduce GHG emissions by means of various management activities that depend on the characteristics of the project site and management goals. For example, the project developer may implement practices that reduce GHG emissions from disturbances through insect and fire prevention or control, or enhance carbon storage by planting improved or different species of planting stock.

3.2 Identifying Primary Effects

Primary effects are the intended changes in GHG emissions or removals associated with a GHG source or sink caused by the project activity. The principal primary effect category for LULUCF project activities is an increase in both CO_2 removals and carbon storage on the project site by means of biological processes, particularly in soil and vegetation.

To estimate the magnitude of the primary effect, the carbon sequestered and stored in or emitted from all carbon pools on the project site should be considered. Potential carbon pools include living biomass, dead organic matter, and soils.

All carbon pools should be included unless the project developer can demonstrate that a pool will not become a source as a result of the project activity. Any aspects of the project implementation that affect these carbon pools should be considered when calculating the magnitude of the primary effect. Chapter 9 offers a more detailed discussion and a list of references for carbon pools and how to measure or estimate them.

TABLE 3.1 Potential Secondary Effects of the Project Activity in Reforestation and Forest Management Project Activities

POSSIBLE PROJECT IMPLEMENTATION ACTIONS CAUSING SECONDARY EFFECT(S)	TYPE OF SECONDARY EFFECT	POTENTIAL GHG EMISSIONS
Reforestation and Forest Management Project Activities		
Fertilization, e.g., on-site fertilizer applications to establish trees and/or promote tree growth.	One-time effect or recurrent upstream effect.	• N_2O from fertilizer. • CO_2 from fossil fuel use during application. • CO_2 from fertilizer manufacture.
Herbicide application, e.g., on-site application of herbicides to remove unwanted vegetation during site preparation or to control weeds during growing period.[B]	One-time effect or recurrent upstream effect.	• CO_2 from fossil fuel use during application. • CO_2 from herbicide manufacture.
Silviculture, e.g., on-site thinning or pruning of trees.[B]	One-time effect or recurrent upstream effect.	• CO_2 from fossil fuel use during thinning or pruning.
Harvest, e.g., on-site removal of trees at end of harvest rotation.[B]	One-time effect or recurrent downstream effect.	• CO_2 from fossil fuel use during harvest.
Transportation, e.g., on- or off-site transportation of products, employees, and inputs such as fertilizer.	Upstream effect.	• CO_2 from fossil fuel use.
Insect or fire control, e.g., improved on-site (and perhaps off-site) practices to reduce insect infestation and fire risk.	One-time effect or recurrent downstream effect.	• CO_2 from fossil fuel use during pesticide application or implementation of fire retardation techniques. • CO_2 from pesticide manufacture.
Replanting, e.g., on-site replanting of trees after harvest.	One-time effect or recurrent upstream effect.	• CO_2 from fossil fuel use during replanting. (Note: fertilizer applications, etc., are included elsewhere.)
Reforestation Project Activities Only		
Site preparation, e.g., on-site mechanical clearing of vegetation and planting preparation.[B]	One-time effect or recurrent upstream effect.	• CO_2 from fossil fuel use.
Nursery production of seedlings.	Upstream effect.	• CO_2 from fossil fuel use. • N_2O from fertilizer.
Forest Management Project Activities Only		
Improved harvest practices, e.g., selective harvesting practices.	One-time effect or recurrent upstream and downstream effect.	• CO_2 from fossil fuel use.
Reduced/selective logging, e.g., on-site increase in length of harvest rotation or selectively logging bigger trees while trying to reduce destruction of adjacent trees.[B]	Downstream effect.	• CO_2 from fossil fuel use during logging.
Cessation of logging, e.g., on-site decision to stop logging activities.[B]	Downstream effect.	• CO_2 from fossil fuel use.

[A] The market responses listed are responses to the secondary effects, not the primary effect. Therefore, they are more likely to be small and may be insignificant.

[B] Project developers should make sure that any carbon emitted or stored through biological process as a result of this management activity—for example, CO_2 and CH_4 from decaying vegetation or soil disturbance—is captured in the primary effect.

16

POTENTIAL MARKET RESPONSES[a] (either positive or negative)
Response to increased fertilizer use. Response to increased fossil fuel use.
Response to increased fossil fuel use. Response to increased herbicide use.
Response to increased fossil fuel use.
Response to increased fossil fuel use.
Response to increased fossil fuel use.
Response to increased fossil fuel use.
Response to increased fossil fuel use.
Response to increased fossil fuel use.
Response to increased fossil fuel use. Response to increased fertilizer use.
Response to increased fossil fuel use. Response to change in fiber supply.
Response to decreased fossil fuel use. Response to change in fiber supply.
Response to decreased fossil fuel use. Response to change in fiber supply.

3.3 Considering All the Secondary Effects

Project activities, or the actions required to achieve the project's GHG goals, often change GHG emissions and sometimes removals beyond the project activity's primary effects. These are the secondary effects and are unintended from the perspective of the GHG reductions, although they may be an integral part of the project activity(ies). For LULUCF projects, secondary effects are primarily changes in nonbiological emissions and biological changes in carbon stocks resulting from market responses (discussed in section 3.3.2 of this chapter).

Secondary effects are usually small compared with the primary effect (see the example in annex A), but occasionally they are large or numerous enough to minimize the intended GHG reductions of the GHG project, thereby rendering the project activity unviable as a GHG reduction effort. For this reason, project developers should consider secondary effects and their possible magnitude before moving forward with the GHG project. The GHG assessment boundary should include all significant secondary effects.

Secondary effects may occur on and off the project site. On-site secondary effects such as the nonbiological GHG emissions from site preparation, planting, fertilizer and herbicide application, and silvicultural activities such as thinning, pruning, or harvesting (e.g., mobile combustion emissions) are usually the easiest to recognize and quantify. These activities are often an integral component of the establishment and maintenance of LULUCF project activities, and they generally can be controlled or influenced by the project developer. An example of secondary effects that may occur off the project site is the GHG emissions from transporting harvested trees.

The Project Protocol divides secondary effects into one-time effects and upstream and downstream effects.

3.3.1 ONE-TIME EFFECTS
One-time effects are secondary effects that are considered only once during the project's lifetime. They are usually related to the nonbiological GHG emissions during the establishment or termination of the project activity, for example, the mobile combustion emissions from land-clearing equipment. Occasionally, one-time effects from LULUCF project activities occur during the life of a project, for example, the mobile combustion emissions from a single precommercial thinning of a forest stand. Table 3.1 shows some examples of one-time secondary effects from reforestation and forest management project activities.

17

3.3.2 UPSTREAM AND DOWNSTREAM EFFECTS

Upstream and downstream effects are recurrent and should be considered throughout the GHG project's operating lifetime. They are related to either the inputs used (upstream effects), like fertilizer, or what is produced (downstream effects) by the project activity. Whether the secondary effect is an upstream or downstream effect depends on the project activity. Table 3.1 also shows some examples of upstream and downstream secondary effects for reforestation and forest management project activities.

Identification of upstream and downstream effects does not require a full life-cycle assessment. Many secondary effects that would be included in a life-cycle assessment would be excluded from the GHG assessment boundary because the size of their GHG emissions, compared with those of other secondary effects or the primary effect, is small (see sections 3.4 and 3.5 on estimating secondary effects and assessing their significance). Annex A provides an example of a life-cycle assessment and the subsequent identification of upstream and downstream effects.

Upstream and Downstream Effects Involving Market Responses

Upstream and downstream effects may provoke responses from the market when alternative producers or the consumers of an input or product react to a change in market supply or demand that is caused by the project activity and results in a change in GHG emissions outside the project site. Market responses vary by location, the nature of the inputs being consumed or the product resulting from the project activity, and specific attributes of the market within which the project activity is operating. Table 3.1 lists some potential market responses for reforestation and forest management projects.

Most of the recognized market responses to reforestation and forest management project activities result directly from the project activity itself, that is, an increase or decrease in the supply of fiber to a local, regional, or global market. The likelihood of market responses depends on

- The extent that the products or services either consumed or produced by the project activity can be replaced by substitutes (e.g., how many other types of building products are available that can be substituted for high-grade building lumber for framing).

- The ability of alternative producers to change their supply of a product or service (e.g., how easily and quickly forestry companies can change their supply of fiber to the market).

- The cumulative impact of similar project activities (e.g., how many other similar GHG projects are in the geographic area and, in aggregate, how these GHG projects will change the supply of a given product or service such as high-grade building lumber).

3.3.3 MITIGATING SECONDARY EFFECTS

Many of the secondary effects of LULUCF project activities result from actions necessary to implement the GHG project. The challenge of LULUCF project activities is mitigating negative secondary effects where possible while recognizing that they probably are an inherent part of the project activity. Mitigating negative secondary effects may reduce their significance enough to exclude them from the GHG assessment boundary (see section 3.5).

TABLE 3.2 Mitigation Options for Secondary Effects from Forest Management Project Activities

PROJECT ACTIVITY	POSSIBLE SECONDARY EFFECTS	MITIGATION OPTIONS
Insect or fire control, e.g., suppression of insect or fire disturbance.	Increase in GHG emissions if additional equipment is needed or the area where suppression practices are implemented is expanded.	• Use fuel-efficient equipment. • Maximize equipment use and effectiveness.
Silviculture, e.g., precommercial or commercial thinning.	Increase in GHG emissions from increased machinery use.	• Use fuel-efficient machinery.
Fertilization.	Increase in N_2O emissions from fertilizer.	• Avoid using volatile forms of ammonium-N, and substitute with urea or other ammonium compounds (ammonium nitrate, ammonium phosphate, etc.).
Enrichment planting or postfire planting.	Increase in GHG emissions from nursery construction and operations.	• Use nitrogen-fixing plants as a cover crop. • Use fuel-efficient machinery and renewable energy for nursery construction and operation.

Reforestation Project Activities

The largest secondary effects for reforestation project activities are usually those associated with the initial land clearing and biomass burning. Because the machinery used to remove vegetation and prepare soil emits GHGs from the fossil fuels it uses, a good way of reducing the magnitude of this secondary effect is to use fuel-efficient machinery.

GHG emissions from fertilizer applications can be reduced by avoiding the more volatile forms of ammonium-N and substituting formulations such as urea or other ammonium compounds (ammonium nitrate, ammonium phosphate, etc.).

Forest Management Project Activities

Secondary effects are often more subtle in forest management project activities because the project activities frequently are a modification of existing practices rather than a new practice. Table 3.2 lists some forest management project activities, their possible secondary effects, and some options for mitigation.

BOX 3.1 Market Responses and GHG Programs

Market responses are the most difficult type of secondary effect to estimate, as they typically extend far beyond the project site. Therefore, complex modeling efforts often are needed to discover how changes in supply or demand affect larger markets. In many instances, GHG programs may be in the best position to estimate these market responses. This would reduce the burden on project developers to estimate these effects, promote consistency in how these market responses are estimated by participants in a given GHG program, and ensure that market responses are routinely considered in any accounting for GHG reductions from LULUCF project activities. For more information about market responses, see the additional references for this chapter in part IV.

Mitigating Market Responses

GHG projects can be designed to reduce or avoid negative market responses. Examples of such design elements are:

- Providing other income streams to displaced workers. For example, land-use GHG projects can accommodate the need for sustainable income by previous users of the land through other economic development efforts, such as ecotourism.

- Offering substitutes for the products or services reduced by the project activity. For example, a forest management project that reduces harvesting activities may fulfill the market demand for fiber by including a forest plantation as an additional GHG project component. Another alternative would be to restock previously understocked forest areas to increase the available supply of fiber.

- Reducing the demand for land by improving its productivity. For example, the demand for agricultural land could be reduced by increasing the intensity with which other land is farmed.

If negative market responses cannot be eliminated or mitigated by the project's design, their possible significance should be determined. If it is not feasible to estimate the market response, this should be clearly documented and explained. If it is estimated, the market response should be factored into the estimation and final quantification of secondary effects.

3.4   Estimating the Relative Magnitude of Secondary Effects

Using Default or Existing Data

Because most secondary effects are small in comparison to the primary effect, using default or existing data may be the most cost-effective approach. The availability of data depends on both the project type and the region where the project is being carried out. One source of information is the supply of life-cycle assessments already made for various types of LULUCF project activities (for an example, see annex A). Often the estimates in these life-cycle assessments can be extrapolated to other regions and, in some instances, to other project types. Another source is documentation from other GHG projects with similar project activities carried out

in various initiatives or GHG programs with publicly available documents. In addition, academic studies or journal articles may have relevant information, for example, in soil conservation and forestry publications.

Using Emission Factors

Emission factors can be used when the secondary effect can be estimated as the product of the emissions rate and the amount of an input used or the resulting product. Many of the secondary effects fall into this category. For example, among the many sources of information pertaining to emissions from various fuel types is the IPCC (see the 2006 IPCC Guidelines for National Greenhouse Gas Inventories) or the GHG Protocol's mobile combustion and pulp and paper calculation tools (www.ghgprotocol.org). These emissions factors can be used along with the input use or production levels to estimate the GHG emissions associated with the fuel used for the GHG project.

The IPCC (2003 and 2006) also provides standardized equations and emission factors to calculate the emissions of nitrous oxide (N_2O) from using fertilizer or of methane (CH_4) from burning biomass. These are developed for estimating national GHG inventories. But if country-specific calculations, which may be available in the country's national inventory report to the UNFCC, are used, they can be very useful. If the country-specific information is not available, the general IPCC equations may still be acceptable.

Undertaking a Market Assessment

A market is assessed in order to determine the size of its response to any changes in supply or demand caused by the GHG project. Various models can be used to estimate the degree of market response,[2] although they may be expensive to use or adapt, and the results tend to vary depending on the assumptions and parameters used. Such models are generally referred to as timber market models or agricultural market models, and they may be national, regional, or global in scale.

Whether a market assessment is necessary depends on the size of the GHG project. Small GHG projects resulting in small changes in supply or demand may not cause an appreciable market response. But if many small GHG projects of the same type are undertaken in one region, then together they may provoke a market response. The estimate of market responses may be best addressed programwide or systemwide by the program or system administrator (see box 3.1). However, absent such a program, project developers will need to consider whether or not to include market responses.

Reforestation Project Activities

Reforestation project activities are likely to cause a market response, as they may substitute one product for another; for example, an agricultural commodity is no longer being produced (decrease in supply), but fiber products are being produced (increase in supply). Determining the amount of the agricultural commodity that is no longer being produced and comparing it with the overall production of the commodity in the region indicates whether a market assessment of the agricultural commodity is needed. If the project displaces only a small percentage of the commodity grown in a region, then a market assessment may not be necessary if an existing oversupply can be verified. A monitoring plan tracking the production of the agricultural commodity over time also should show whether the loss in production resulting from the GHG project is now being compensated by other producers in the region.

Similarly, the increase in fiber may cause a market response if less fiber is being harvested elsewhere. This, however, is a positive secondary effect and can be excluded if a market assessment is too expensive (see section 3.5 on assessing the significance of secondary effects).

Forest Management Project Activities

As with reforestation project activities, estimating the increase or decrease in fiber resulting from a project activity and comparing it with the region's supply of fiber indicates whether a market response may be expected. The monitoring of fiber production also should show whether fiber production has changed over time. If it has not changed, then some producers have produced less fiber and provoked a market response that may need to be taken into account.

Applying the Conservativeness Principle

Uncertainty is intrinsic to any estimate of secondary effects. Accordingly, the conservativeness principle should guide any effort to estimate their magnitude. For instance, it is advisable to use upper-bound estimates for a project activity's nonbiological GHG emissions and compare them with lower-bound or zero estimates for baseline nonbiological GHG emissions. Similarly, if a market response is possible, the project developer should be conservative in determining whether it is likely to occur. In some instances, being conservative may mean assuming that the entire supply of a product or the entire demand for an input has been substituted.

3.5 Assessing the Significance of Secondary Effects

The significance of a secondary effect is usually determined by comparing the size of the secondary effect with the size of the primary effect. To some extent, this is a subjective judgment, but the following criteria, listed in the Project Protocol, may help establish when a secondary effect is not significant:

- The secondary effect makes a positive or no difference between the baseline emissions and the project activity emissions. Some actions associated with the primary effect from the LULUCF project cause GHG emissions. However, if these actions decrease the GHG emissions compared to baseline emissions, these are considered positive secondary effects and can be excluded if desired. Likewise, when the project activity and the baseline emissions from a given source do not change, there are no secondary effects, and so they do not need to be included in the GHG reduction calculation. For instance, if nitrogen fertilizer is required in the baseline scenario and for the project activity, but the quantities in both cases are similar, those GHG emissions can be excluded.

- The secondary effect is small relative to the associated primary effect. When the secondary effect is small, its exclusion may be justified. To make this comparison, both the secondary effect and the primary effect must be estimated. Generally, the further upstream or downstream the effect is, the less likely the secondary effect it is to be significant.

- The secondary effect has a negligible market response. In some instances, the expected market response is small or negligible, usually when the change in supply or demand of a product is small compared with the size of the market for that product. In this case, the market response may be considered insignificant.

NOTES

[1] For the definitions of forest, afforestation, reforestation, etc., see the specific GHG program guidelines. Annex B also has some examples.

[2] The project developer should ensure that any model used has undergone a peer review or has otherwise been tested by a credible institution.

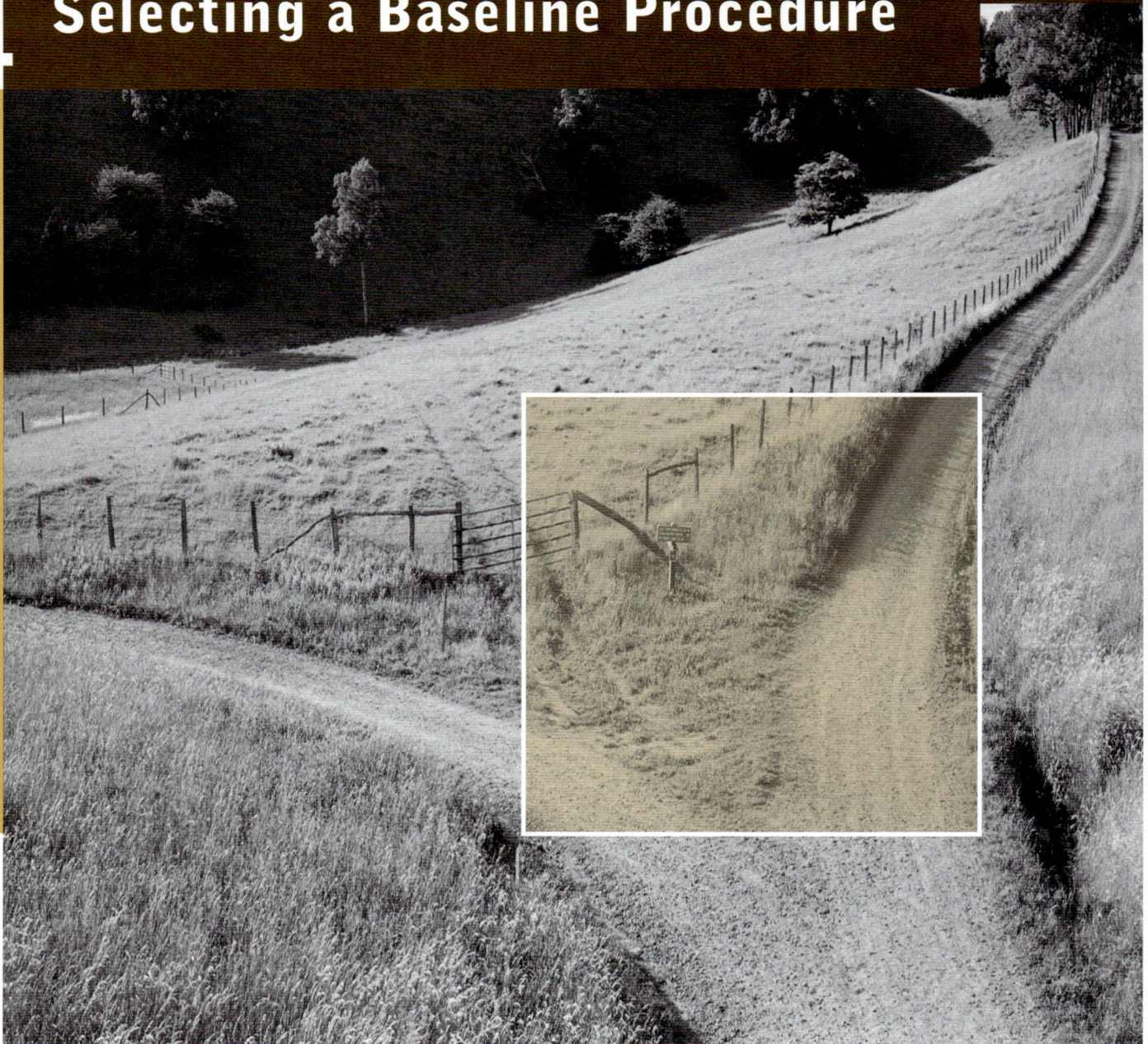

4 Selecting a Baseline Procedure

This chapter introduces the two baseline procedures and offers guidance on selecting either the project-specific or the performance standard procedure.

4.1 Describing the Baseline Procedures

Project-Specific Procedure

The *project-specific procedure* uses the specific circumstances or conditions relating to the project activity to identify a *baseline scenario,* from which the corresponding baseline carbon stocks, change in carbon stocks and GHG removals are estimated. The procedure uses a structured, mainly qualitative, analysis of the project activity and its alternatives to determine which alternative (including the project activity) faces the fewest or least significant barriers to implementation. If that analysis does not provide enough information, then the alternative with the greatest number of net benefits not related to climate change mitigation should be chosen. This alternative is the one most likely to have been implemented without considering climate change objectives. It is then assumed to be the "baseline scenario" from which the baseline GHG removals are calculated and is used for comparison with the GHG removals from the project activity. If the project activity and the baseline scenario have the same GHG removals, the project activity will not have any GHG reductions. The baseline scenario is usually valid only for the specific project activity being examined.

Performance Standard Procedure

The *performance standard procedure* is used to develop a *performance standard*, sometimes referred to as a *multiproject baseline* or *benchmark,* because it can be used to estimate baseline GHG removals for several project activities of the same type. The performance standard serves the same function as a baseline scenario in the project-specific procedure but avoids the need to identify an explicit baseline scenario for each project activity. Instead, a quantitative analysis of the baseline candidate GHG removals is used to define baseline GHG removals that can be used by different project developers for the same project activity within a specified region.

4.2 Selecting a Baseline Procedure

The Project Protocol has four criteria for selecting a baseline procedure, whose applicability to LULUCF projects is explored in this section.

1. **The number of similar project activities that may be implemented across a given area.**
   When a number of GHG projects with similar project activities are in the same geographic area, the performance standard is an attractive option. But when only a single GHG project is being considered, the project-specific procedure may be better.

2. **The ability to obtain verifiable information about implementing baseline candidates.**
   Although this consideration can affect the collection of information for either of the procedures (e.g., information about the management practices used by different landowners is needed for both procedures), verifiable information about the barriers or benefits of the practices (e.g., finance-related barriers or benefits) used by other landowners may be difficult to obtain and be considered confidential. In such cases, the project-specific procedure may be more difficult to apply, especially to forest management project activities.

3. **Confidentiality concerns regarding information about the project activity's costs and benefits.**
   Forest management project activities are the most likely to face confidentiality concerns about business information when different business practices are discussed, which may make the project-specific procedure more difficult to apply.

4. **The number of baseline candidates or the availability of carbon stock and GHG removal data for each baseline candidate.**
   In the LULUCF context, the "number" of baseline candidates for the performance standard is the number of units of land area representing a land-use or management practice in the defined geographic area. The GHG data for each baseline candidate (e.g., each hectare) can be difficult to obtain, however, which can complicate the application of the performance standard. If reliable GHG data can be obtained for a representative number of sites for each type of baseline candidate and credibly extrapolated to the entire geographic area, then the performance standard can be used. In both the project-specific and performance standard procedures, accurate data on baseline carbon stocks and GHG removals are necessary.

5 Identifying the Baseline Candidates

The baseline candidates for LULUCF project activities are the alternative land uses or management practices on lands in a specific geographic area and during a given temporal range.

This chapter closely follows chapter 7 of the Project Protocol and helps the project developer compile a final list of baseline candidates to be used in the baseline procedures (see box 5.1). This list should:

1. Consider the product or service provided by the project activity.

2. Consider the types of applicable baseline candidates.

3. Refine the list of baseline candidates by defining a geographic area that reflects the characteristics of the project site and a temporal range that helps show any trends and changes in the geographic area.

4. Consider any other characteristics that may be important.

> **BOX 5.1 GHG Programs and Baseline Candidates**
>
> Drawing up a list of baseline candidates is one of the best ways of ensuring the environmental integrity of both baseline procedures. It is also one of the more time-consuming and data-intensive steps in the Project Protocol. When possible, GHG programs that include GHG projects using either the performance standard or the project-specific procedure to estimate baseline GHG removals could provide the list of baseline candidates that the project developer would use for a particular project type in a specific geographic area. Such a list would also make specific project activities more attractive by relieving the project developer of some of the data collection burdens and uncertainty regarding which baseline candidates should be considered.

5.1 Defining the Product or Service Provided by the Project Activity

A list of possible types of baseline candidates should contain all activities that could offer identical (or nearly identical) products or services as the project activity. For some LULUCF projects, however, the baseline candidates and the project activity do not provide the same product or service, and so it is not necessary to define the product or service when identifying possible baseline candidates. When the product or service is not relevant to making an initial list of baseline candidates, determine the alternative practices or land uses that could be undertaken on the project site (see section 5.2).

Reforestation Project Activities

The primary effect for reforestation project activities results from a change in land use, and the project activity and the baseline candidates most likely provide a different product or service. For instance, the product of a reforestation project may be wood fiber, whereas the products provided by the baseline candidates may range from animal forage from pastureland to commercial grains from a corn/soybean cropping rotation.

Forest Management Project Activities

The baseline candidates for forest management project activities generally have the same type and/or quantity of a particular product, such as x board feet of lumber. The following questions may help determine whether specifying the type and/or a quantity of product or service is useful for identifying the baseline candidates:

- Will the project activity change the current type of forest management? If yes, then specifying both the type and quantity of product being provided may help identify the baseline candidates.

- Will the project activity change the intensity of the current forest management? If yes, then the type of product being provided will help identify the baseline candidates, but quantity will not.

- Will the project activity change forest management practices pertaining to a specific type and quality of product or service being provided? If so, the type and quality of the product provided may help identify the baseline candidates. For example, if the project activity is to change a management practice specifically for hardwood fruit trees used for furniture-quality lumber, then management practices related to creating furniture-quality lumber may be used to define the other baseline candidates.

5.2 Identifying Possible Types of Baseline Candidates

When the products or services of the project activity and/or the baseline candidates are different, possible baseline candidates are identified by looking at sites similar to the project site and identifying the land uses or management practices taking place or being planned on these sites.

Reforestation Project Activities

Some questions that may help identify the types of baseline candidates for reforestation project activities are:
- What are the current land uses on land similar to the project site?

- What are the possible alternative land uses on land similar to the project site?

Possible baseline candidates for reforestation project activities are:
- Commercial crop production, which may be further divided into cereal, grain, fiber, vegetable, etc. Note whether single crops or cropping rotations have been implemented.

- Permanent subsistence crop production, which may be further divided into types of crops grown under subsistence production.

25

- Pastureland, which may be further divided into range or managed pasture, improved or unimproved pasture, irrigated or nonirrigated pasture, permanently or rotationally grazed, and so forth.

- Abandoned land.

- Unused, degraded, or unproductive land, for example, areas with severely eroded soils or areas disturbed by mining.

- Forest, which may be further divided into riparian buffers, various types of plantation forests, naturally regenerated forests, native or exotic forests, forests composed of different tree species, and so on.

- Developed land, which may be further divided into hobby farm/lifestyle blocks, residential (single or multifamily), commercial, or industrial development.

Forest Management Project Activities

Forest management project activities change either the intensity or the type of forest management being practiced. When the project activity changes the intensity, the baseline candidates will most likely represent the management practices' various intensities. For example, if the project activity changes the length of the harvest rotation, then the baseline candidates will represent all the possible harvest rotations.

If the project activity changes the type of forest management, then the baseline candidates are types of forest management practices that provide the same service. For instance, mechanical weeding, chemical weeding, and thinning forest management practices are comparable, as they all perform the same service: reducing the plant density and ensuring the survival of desirable trees. In contrast, site preparation practices should not be compared with harvest practices, since the first tries to improve forest regeneration and the second tries to increase timber volume.

Some questions that may help identify baseline candidates for forest management project activities when the project activity and baseline candidates offer an identical (or nearly identical) product or service are:

- What alternative new or existing forest management practices would produce a similar product or service as the project activity on the project site?

- What alternative new or existing forest management practices do others use to produce a similar product or service for the project activity?

- What is the most common forest management practice on sites that produce the same (or similar) product or service as the project activity in the same market?

- What is the existing forest management practice?

5.3 Defining the Geographic Area and the Temporal Range

5.3.1 DEFINING THE GEOGRAPHIC AREA

The geographic area is the location of the land uses or management practices of the final list of baseline candidates. This area should be larger than the project site and could include lands spanning a state or province, a group of states or provinces, or maybe even a country. The geographic area does not need to be contiguous but includes only lands similar to the project site. It is useful to think about the project site's inherent production capabilities. For instance, if the project site is suited to dryland pasture but not dryland cereal production, then dryland cereal production areas should not be included in the geographic area. When defining the geographic area in order to narrow the list of possible types of baseline candidates, consider the following constraints:

- Biophysical conditions that reflect the characteristics of the project site and influence both what type of plant species could be grown and how well they grow, for example, climatic conditions (like precipitation and temperature) or geological conditions (like soil type and topography).

- Human-influenced factors that may affect what plant species would be grown, for example, legal, sociocultural, or economic factors.

- Availability of physical infrastructure, for example, roads or markets.

There is no hard or fast rule for which criteria are most important to defining the geographic area in a LULUCF context or for the order in which they should be applied. But applying the criteria in the following order generally helps define the geographic area:

1. **Political boundaries:** National or subnational boundaries often are the first consideration for defining the geographic area. Many land-use regulations, government incentive programs, and the like are specified according to political boundaries, and these legal requirements or programs often affect land-use or management practices.

2. **Ecological zone:** Many countries have classified ecological zones that have been determined by government agencies. For areas for which government classifications are not available, environmental organizations like the Nature Conservancy or the World Wildlife Fund may have mapped ecological zones. For many LULUCF project activities, the geographic area will be constrained by an ecological zone.

3. **Land characteristics:** More specific land characteristics, such as particular soils, topography, or proximity to rivers and streams, often can be used to narrow the geographic area.

4 **Land-use/management history:** Past land use or management can influence the land's productivity. Therefore, the geographic area should probably be restricted to areas with a similar land-use or management history.

5. **Other factors:** Other social, cultural, or economic factors, such as market accessibility and infrastructure availability, may be important to defining the geographic area. For instance, grain-handling depots and rail lines may be necessary to transport cereal grain from where it is grown to larger markets or seaports. Therefore, limiting the geographic area to where these depots and rail lines are located may make sense. Another example is certain areas may be regarded as sacred by indigenous populations, meaning that certain land uses or management practices are prohibited and thus should be excluded from the geographic area. In other cases, the GHG project may service a specific market, such as fiber grown for ethanol production, and so its proximity to an ethanol plant may be a factor limiting the geographic area.

When defining the geographic area for LULUCF project activities, remember that
- The geographic area does not need to be contiguous.

- The baseline candidates may need to be further refined/adjusted at a later stage according to the availability of land-use or management data or carbon data.

Reforestation Project Activities
Biophysical conditions are one of the determining criteria for the geographic area of reforestation project activities, as these conditions typically define the type of vegetation, and therefore the land use suitable to an area. Regional zoning regulations or conservation incentive programs are examples of legal factors that may affect the geographic area for these activities. The importance of other characteristics is likely to differ by region and should be assessed on a case-by-case basis.

Forest Management Project Activities
Many countries already have ecological classification systems describing the conditions for forest management. When these areas have not been identified, a coarse-scale classification can be created from air photos or satellite images. Forest inventory systems are often stratified by type of forest cover or other vegetative characteristics and can also be the basis for defining the biophysical aspects of the geographic area for forest management project activities. However these inventories often are available only in areas of commercial forestry.

Many areas have forest management regulations that also can guide the definition of the geographic area. For instance, California has quite comprehensive forest management laws compared with those of other U.S. states, so if the project site is to be located in California, it would make sense to restrict the geographic area to that state or even a smaller area within the state.

As with reforestation project activities, other characteristics are likely to be specific to individual regions and so should be assessed as such.

5.3.2 DEFINING THE TEMPORAL RANGE
The temporal range is the time period used to determine the types of baseline candidates (see box 5.2). The temporal range for LULUCF project activities tends to be longer than that of most emission reduction project activities.

BOX 5.2 Identifying Baseline Candidates Using the Temporal Range

Project developers may be able to exclude certain baseline candidates if they can show that no lands under the same historic land-use or management practice as the project site have ever changed to that particular baseline candidate within the temporal range.

For example, a project developer has identified four baseline candidates after defining the geographic area: cropland (the project site's historical use), pastureland, abandoned land, and land being developed for residential uses. After looking at two geographic information system (GIS) datasets from the beginning and the end of the temporal range, the project developer finds that during the temporal range, the cropland in the geographic area has never been abandoned and so abandoned land is not a baseline candidate. However, if the project developer does not have data to determine specific land-use or management changes for each land unit or cannot find the data for a sufficiently long or representative temporal range, then baseline candidates cannot be excluded using this approach.

Two factors for defining the temporal range for LULUCF project activities are
- An event that clearly marks a "discrete point" in time when management and land-use decisions changed. The timing of that event represents the beginning of the temporal range. For example, after a severe flood, if the landowners switched from agricultural crops to short-rotation forests on river floodplains, then the flood would mark the beginning of the temporal range. Similarly, if regulations were introduced that required certain changes in forest management practices, then the introduction of the regulation would be the beginning of the temporal range.

Regardless of the criteria used to define the geographic area and temporal range, make sure to present all the criteria and information clearly and transparently.

5.4 Defining Other Criteria Used to Identify the Types of Baseline Candidates

Common practice and legal requirements are two other factors that usually can be used to refine the list of baseline candidates. Other criteria may be important as well and should be clearly and transparently documented.

Reforestation Project Activities

Legal requirements affecting land use may be quite diverse, such as protecting waterways and restricting land use on different slope classes. These legal requirements can be used to narrow the list of baseline candidates. For example, annual cropping may not be permitted on land with a slope greater than 8 percent. Therefore, if the project site has such a slope, annual cropping would be excluded as a possible baseline candidate.

Forest Management Project Activities

The forest management legislation for many regions either prescribes or requires certain activities (e.g., reforestation is required after harvesting). These regulations, which may have been used to eliminate regions from the geographic area, can also be used to exclude those baseline candidates that do not satisfy them. For example, replanting after harvest is a forest management requirement in California, U.S., so any baseline candidate that does not meet this requirement would be excluded from a forest management project in this state.

As always, when taking into account legal requirements, the level of enforcement should be considered. For more information, see annex A of the Project Protocol.

- Trends or patterns in production cycles or land-use changes. The temporal range should include at least one full cycle of the longest trend or pattern. For example, if the baseline candidates represent different harvesting rotations, the longest rotation should define the temporal range, or if agricultural land-use decisions are cyclical, reflecting changing commodity prices, then the temporal range would reflect one cycle of commodity prices changes (i.e., from the high prices through the low prices to the high prices again).

If there is no trend or pattern in production cycles or land-use changes and no distinct events marking a change in land-use or management practices, then a temporal range of at least ten years should be used. If reliable data that far back are not available, then the temporal range should be based on the available data, with some explanation of why less than ten years was used (see box 5.3).

5.5 Identifying the Final List of Baseline Candidates

Once the geographic area, temporal range, and any other criteria for identifying the baseline candidates have been decided, draw up the final list of baseline candidates. The land uses or management practices in the list define the final geographic area and temporal range. In addition, for the performance standard procedure, besides the baseline candidates the list should specify the individual units of land area (e.g., hectares) for each land-use or management practice in the geographic area.

BOX 5.3 Multiple Changes in Land-Use or Management Practices on the Same Land Area over Time

Within a given temporal range, land-use or management practices on certain areas of land may have changed many times. Depending on the nature of these changes, it may be necessary to redefine the baseline candidate or extend the temporal range to find out whether these changes show a consistent pattern. The baseline candidate should be redefined when:

- The land-use or management changes on a given land area follow a pattern. In this case, the baseline candidate should represent the pattern. For example, if the land use changes regularly between growing corn and soybeans, then the baseline candidate would be a corn/soybean rotation.

- One land-use or management practice predominates on a given land area. In this case, the predominant land-use or management practice at the end of the temporal range should be used. The project developer should also determine where the trend appears to solidify and use this point to reset the beginning of

the temporal range. For example, if the temporal range originally was twenty years but in the tenth year a specific land use began appearing more often than others, the temporal range should be limited to ten years unless other baseline candidates require a longer temporal range.

If it is not possible to redefine the baseline candidates in this manner, the project developer could also redefine the baseline candidates more broadly (e.g., classifying all cropping baseline candidates as cash crops or subsistence crops) to see whether a pattern emerges.[1]

Redefining baseline candidates may require significant data resources, which may not, in all cases, be available. If the project developer cannot find sufficient data, defining the baseline candidates based on the current data is acceptable. Although it may not capture all the fluctuations of land uses over time, most of the appropriate candidates should be captured, given a large enough geographic area.

5.6 Identifying Types of Baseline Candidates That Represent Common Practice

From the final list, choose the baseline candidate(s) that represent what could be considered common practice (for guidance on defining common-practice land uses or practices, see chapter 7 (section 7.4.2) in the Project Protocol). For LULUCF project activities, common practice usually is represented by the predominant land use for reforestation project activities or the predominant forest management practice for forest management project activities in the geographic area. An explanation should accompany any common practice that cannot be defined.

NOTES

[1] Project developers should note that when baseline candidates are merged, the potential variability in the carbon stocks may increase, making the estimate of the GHG removals less accurate.

30

When using the project-specific procedure, estimate baseline GHG removals by identifying the baseline scenario and accounting for the carbon stocks, the change in carbon stocks and the GHG removals associated with that baseline scenario. Chapter 8 and annex C of the Project Protocol describe in detail the components of the project-specific procedure. The following guidance offers supplementary information for LULUCF project activities, particularly reforestation and forest management. This chapter is best used in conjunction with chapter 8 of the Project Protocol.

The three types of possible alternatives for the baseline scenario are

- The same practice(s) or land use(s) as those represented by the project activity.

- The baseline candidates identified in chapter 5.

- The continuation of current activities, which also should have been singled out in chapter 5 as one of the baseline candidates. This is the project site's existing land use for reforestation project activities or the current management practice for forest management project activities. The continuation of current activities is separated out because the barriers facing the continuation of current activities will most likely differ from those facing the project activity and other baseline candidates.

6.1 Performing a Comparative Assessment of Barriers

6.1.1 IDENTIFYING BARRIERS TO THE PROJECT ACTIVITY AND BASELINE CANDIDATES

Barriers should include anything that would discourage the implementation of the project activity or baseline candidates. The project activity and baseline candidates each may face multiple barriers. Table 6.1 lists the major categories of possible barriers (along with examples). When identifying barriers, examine each barrier category and explain how each would affect each baseline candidate.

6.1.2 IDENTIFYING BARRIERS TO THE CONTINUATION OF CURRENT ACTIVITIES

In most cases, there are no barriers to the continuation of current activities, but if there are, they often are prohibitive or insurmountable. Some barriers for reforestation and forest management project activities are

- Legal or regulatory changes, for example, certain management practices or land uses required in specific areas.

- Unfavorable public perception, for example, certain forest practices' increased risk of fires or dislike of genetically modified organisms (GMOs) grown on agricultural lands.

- Land degradation, for example, current activities' deterioration of the land, through high soil loss and lowered soil fertility, to an extent that a change in land use is required.

- Climate/environmental challenges, for example, the depletion of underground aquifers used for irrigation water, resulting in land-use and/or management changes using less water; high weather variability making the area no longer suitable for certain land uses or management regimes.

6.1.3 ASSESSING THE RELATIVE IMPORTANCE OF THE IDENTIFIED BARRIERS

Barriers are rarely absolute in that they eliminate an alternative from further consideration. To compare the different alternatives, assess the relative importance of each barrier and then the relative importance of each barrier to each alternative.

6.2 Identifying the Baseline Scenario

6.2.1 EXPLAINING BARRIERS TO THE PROJECT ACTIVITY AND HOW THEY WILL BE OVERCOME

A project activity usually faces at least one barrier. For it to proceed, specific measures or design features are added to overcome these barriers and should be documented.

Examples of such measures and/or design features that may be relevant to reforestation or forest management project activities are those that:

- Contribute to the transfer of new technologies or practices, for example, training for new forest management techniques or silviculture practices that enhance forest growth or carbon storage.

- Enter into partnerships with other landowners/timber companies/governments to support the construction of new infrastructure, for example, roads, mills, or nurseries.

- Introduce innovative financing arrangements that offset the risks associated with high up-front costs and delayed revenue streams.

- Offer campaigns to promote new forest management techniques or increase support for land-use changes, for example, from an environmental sustainability perspective.

31

TABLE 6.1: Categories and Examples of Barriers

TYPES	EXAMPLES

Financial and budgetary	**General**
	• High investment costs (e.g., new equipment, infrastructure, management practices).
	• Limited or no access to capital.
	• High perceived risks, resulting in high borrowing costs or lack of access to credit or capital. Perceived risks might be associated with, among other things,
	   • disturbances (e.g., fire, disease),
	   • political instability,
	   • currency fluctuations,
	   • regulatory uncertainty,
	   • poor credit rating of project partners,
	   • unproven technologies, practices, or business models,
	   • general risk of project failure.
	
	Reforestation
	• Long lag time between up-front costs and revenue stream for forest projects.
	
	Forest Management
	• Low marginal returns on investment for additional management effort.

Technology operation and maintenance	**General** • Lack of trained personnel capable of maintaining, operating, or managing a new land use (e.g., forest) and lack of education or training resources to learn the required management skills.
Infrastructure	**Reforestation** • Inadequate transport infrastructure for harvesting, e.g., roads. • Lack of infrastructure for processing forest products, e.g., mills. • Lack of infrastructure for planting new forest areas, e.g., nursery stock availability.
Market structure	**Reforestation** • No local or regional market infrastructure for forest products.
Institutional/ social/ cultural/ political	**General** • Lack of consensus on future management decisions (e.g., with respect to land use). • Lack of clear ownership of carbon rights for publicly held land. **Reforestation** • Social and/or cultural ties to land resulting in reticence to change land use. • Poor public perception of certain land uses, e.g., perception of competition for land—trade-off among food, forests, and urban uses—or high water use by forests (e.g., Eucalyptus forests) versus agricultural land. **Forest Management** • Ingrained traditional forest management practices.
Resource availability	**Reforestation** • Lack of sufficient irrigation water for agricultural crops or young forests.

Note: This list is not intended to be exhaustive. Project developers or GHG programs may discover other barriers not described here.

6.2.2 IDENTIFYING THE BASELINE SCENARIO USING THE COMPARATIVE ASSESSMENT OF BARRIERS

In many instances the baseline scenario can be identified as the candidate facing the fewest barriers. This decision should be justified using a comparative assessment of barriers. In those cases when no clear baseline scenario emerges from the barrier assessment, there are two options: (1) using the most conservative viable alternative or (2) conducting a net benefits assessment of the remaining baseline candidates.

A *net benefits assessment* can be a quantitative or qualitative assessment and include the
- Expected financial returns (assessed either qualitatively or quantitatively).

- Public relations benefits, for example, improving or maintaining the landscape's aesthetic beauty.

- Research and demonstration value for new management practices.

- Position or entry in a specific market, strategic alignment, or other competitive reasons.

- Environmental benefits, for example, improved biodiversity, water quality, soil fertility, habitat, or reduced soil loss.

These benefits should be measured from the perspective of the project developer, not society in general.

6.2.3 JUSTIFYING THE BASELINE SCENARIO

The baseline scenario should be the baseline candidate with the fewest barriers and, if applicable, the highest net benefits. Comparing the identified baseline scenario with common practice can also help strengthen the justification of the baseline scenario. The baseline scenario should be documented.

6.3 Estimating the Baseline GHG Removals and Total Carbon Stocks

The baseline GHG removals associated with the identified baseline scenario are estimated by accounting for the change in carbon stocks corresponding to the baseline scenario and then finding the corresponding GHG removals. The baseline change in carbon stocks (t C/ha) are calculated as the difference in the baseline scenario's carbon stocks between two time periods (for an example see table 7.1 in chapter 7). GHG removals (t CO_2/ha) are calculated by multiplying the change in carbon stocks by $\frac{44}{12}$, the ratio of the molecular weight of CO_2 to the molecular weight of carbon. If a land-use trend factor is relevant, this should also be considered when calculating the baseline GHG removals.

In addition, total carbon stocks for the baseline scenario should also be calculated. Total carbon stocks are found by calculating the carbon stocks per hectare at the beginning of the project time period and adding any changes in carbon stocks per hectare expected during the project's lifetime. This total per hectare is then multiplied by the area covered by the project activity to derive the total carbon stocks.

The equations for finding GHG removals and total carbon stocks are in box 6.1. Guidance on accounting for carbon stocks and resources to use can be found in chapter 9, Estimating and Quantifying Carbon Stocks. In addition, calculating changes in carbon stock, GHG removals and total carbon stocks is illustrated in the Nipawin Afforestation Project example.

BOX 6.1 Calculating GHG Removals and Total Baseline Carbon Stocks

GHG Removals

Baseline Change in Carbon Stocks$_{pzt}$
$$= \text{Baseline Carbon Stocks}_{pzt} - \text{Baseline Carbon Stocks}_{pz(t-1)}$$

Where

Baseline Carbon Stocks$_{pzt}$ = Σ baseline carbon stocks from each biological carbon pool measured, k, related to each primary effect, p, for project activity, z, in period t for a given unit of land area.

Baseline GHG Removals$_{pzt}$ = Baseline Change in Carbon Stocks$_{pzt}$ • $\frac{44}{12}$ t CO_2/t C

Total Baseline Carbon Stocks

Total Baseline Carbon Stocks (t C)
$$= (\text{Carbon Stocks at time zero (t C/ha)} + \sum_{t=1}^{t=n} \text{Baseline Change in Carbon Stocks}_{pzt} \text{ (t C/ha)} \cdot \text{Project Site Area (ha)}$$

Where

n = the final time period.

7 Estimating the Baseline GHG Removals— Performance Standard Procedure

The performance standard procedure, discussed in the Project Protocol, also can be used to estimate the baseline GHG removals against which the project activity GHG removals are compared in chapter 10, Monitoring and Quantifying GHG Reductions.

This chapter describes the time-based performance standard procedure for estimating the baseline GHG removals for LULUCF project activities. The production-based performance standard, the other performance standard outlined in the Project Protocol, is not applicable to LULUCF project activities, since the product or service provided by the baseline candidates often is different. In addition, the production-based performance standard does not capture the time and size dimensions of LULUCF project activities. In this chapter, the performance standard's total carbon stocks are also found.

The time-based performance standard procedure follows the steps outlined in chapter 9 of the Project Protocol, and should be used in conjunction with this guidance (also see box 7.1). The steps are as follows:

- Specify the appropriate performance metric.

- Calculate the GHG removals for each baseline candidate in each time period.

- Calculate the GHG removals for different stringency levels.

- Select an appropriate stringency level.

- Estimate the baseline GHG removals and total carbon stocks.

7.1. Time-Based Performance Standard

The time-based performance standard can be used to estimate baseline GHG removals for a specified project type in a given geographic area.

For each baseline candidate, estimate the change in carbon stocks per unit area per time period during the lifetime of the project activity, and using this information, calculate the GHG removals. In each time period, compare the GHG removals for each baseline candidate by looking at the mean, median, and different percentiles of the GHG removals, as well as the most stringent GHG removals.

7.1.1 SPECIFY THE APPROPRIATE PERFORMANCE METRIC

The time-based performance standard metric described in the Project Protocol focuses on GHG removals but does not completely describe the performance metric that should be used for LULUCF projects. For these, the performance metric should be

$$\frac{GHG\ removals_t}{unit\ area\ of\ land}$$

Where GHG removals (t CO_2/ha) are calculated by multiplying the change in carbon stocks by $\frac{44}{12}$, the ratio of the molecular weight of CO_2 to the molecular weight of carbon.

The performance metric is used for LULUCF project activities to define the baseline GHG removals at regular intervals throughout the activity's life. For example, if a project activity's life is twenty years, estimate the baseline tonnes of carbon stored per hectare for each year, or perhaps five years, thereby capturing the dynamic nature of carbon stocks.

The length of each time period for carbon stock estimations varies according to the rate at which different plant species store carbon at a given location.

7.1.2 CALCULATE GHG REMOVALS FOR EACH BASELINE CANDIDATE IN EACH TIME PERIOD

To calculate GHG removals, estimate the baseline change in carbon stocks per time period and unit of land area for each baseline candidate during the lifetime of the project activity. There are two approaches for calculating the change in carbon stocks; the project developer may find information on the carbon stocks over time for a given land use and calculate the difference between two points in time to get the change, or there may be information already available which can be translated into the changes in carbon stocks for a particular activity (e.g., yearly tree growth curves can be translated into yearly carbon stored for certain carbon pools). The data needed to estimate carbon stocks and carbon stock changes may already exist or may need to be collected or estimated specifically for the baseline candidates. Methods of collecting data and estimating baseline carbon include direct measurement, statistical sampling, proxies, modeling, default values, and remote sensing. Project developers most likely will need to use a combination of methods to estimate baseline carbon stocks and carbon stock changes.

A major decision when estimating the carbon stocks changes for LULUCF baseline candidates is which carbon pools to include in the estimation. The carbon associated with each carbon pool assessed is summed to obtain the carbon stocks for each baseline candidate in each time period. The same carbon pools should be considered for each baseline candidate, and if different carbon pools are found to be relevant, this should be noted (see box 7.2). More information about the carbon pools to assess and the available measurement/estimation methods can found be in chapter 9, Estimating and Quantifying Carbon Stocks, along with a list of resources for a more detailed discussion of the different measurement or estimation methods.

35

Whether the carbon stocks and/or changes in carbon stocks are measured directly from the individual baseline candidates (or some subset of these) or derived using some other method, be sure to note the type and degree of variability and uncertainty of these estimates. Similarly, if existing carbon stock information is used, the variability in the values and their level of uncertainty should be noted as well.

Box 7.3 describes the initial forest inventory used to calculate the GHG removals from various management practices.

Baseline GHG Removals

Once the baseline change in carbon stocks per unit of land area and time period have been estimated or measured for each baseline candidate, the GHG removals (t CO_2/ha) are calculated by multiplying the change in carbon stocks by $\frac{44}{12}$, the ratio of the molecular weight of CO_2 to the molecular weight of carbon. See box 6.1 for the equations to calculate the baseline GHG removals and table 7.1 for an example of the calculations.

7.1.3 CALCULATE THE GHG REMOVALS FOR DIFFERENT STRINGENCY LEVELS

The stringency of a time-based performance standard for LULUCF project activities refers to how large the baseline GHG removals are relative to the GHG removals of all the baseline candidates. The stringency level is essentially a better-than-average GHG removal.

TABLE 7.1 Calculating GHG Removals

TIME PERIOD	Baseline Candidate 1			Baseline Candidate 2		
	BASELINE CARBON STOCKS, t C/ha	CHANGE IN CARBON STOCKS, t C/ha	GHG REMOVALS t CO₂/ha	BASELINE CARBON STOCKS, t C/ha	CHANGE IN CARBON STOCKS, t C/ha	GHG REMOVALS t CO₂/ha
0	854	0	0	854	0	0
1	764	-90	-330	852	-2	-7.3
2	754	-10	-36.7	851	-1	-3.7
3	746	-8	-29.3	852	1	3.7
4	739	-7	-25.7	853	1	3.7
5	733	-6	-22	855	2	7.3
6	728	-5	-18.3	857	2	7.3
7	723	-5	-18.3	859	2	7.3
8	720	-3	-11	862	3	11
9	716	-4	-14.7	865	3	11
10	714	-2	-7.3	867	2	7.3

To determine the appropriate stringency level to use (see section 7.1.4), a number of different stringency levels are compared for each time period:

- Most stringent: The baseline candidate with the highest GHG removals for any given time period (this may change over time).

- Weighted mean GHG removals.

- Median GHG removals (50th percentile).

- GHG removals relating to two different percentiles that are better than average (e.g., 75th and 90th percentile).

These stringency levels are calculated for each relevant time period (e.g., every year or every five years) during the life of the project activity. The equations given find the stringency level for only one time period and must be applied to each additional relevant time period. Box 7.4 provides an illustrative example of the different stringency level calculations.

Most Stringent Stringency Level

The *most stringent stringency* level is the baseline candidate(s) with the highest GHG removals for the specified time period. If the baseline candidate with the highest GHG removals changes between time periods, this should be noted.

> **BOX 7.3 Calculating Baseline Carbon Stocks, Carbon Stock Changes, and GHG Removals for Forest Management Projects**
>
> For a forest management project, to determine the baseline carbon stocks changes and GHG removals of the baseline candidates identified in chapter 5, Identifying the Baseline Candidates, apply each representative management practice to the project site's current forest inventory, and model or estimate the resulting carbon stocks and change in carbon stocks. Although this offers a more consistent comparison of how various management practices affect forest carbon storage on the project site, it does affect the stringency level defining the performance standard and is described in section 7.1.4.

Weighted Mean GHG Removals

To calculate the *weighted mean GHG removals* in each time period, use the following equation:

$$\text{Weighted mean GHG removals}_{jt} = \frac{\sum_{j=1}^{n} (\text{CO}_2 \text{ removals}_{jt} \cdot \text{area}_j)}{\sum_{j=1}^{n} (\text{area}_j)}$$

where

$\text{GHG removals}_{jt} = $ GHG removals for baseline candidate j in time period t

$\text{area}_j = $ area encompassed by baseline candidate j (e.g., 1 hectare)

$n = $ total number of baseline candidates

$j = $ individual baseline candidate

Median GHG Removals

To calculate the *median GHG removals,* find the 50th percentile using the percentile method below.

GHG Removals Relating to Different Percentiles

The following approach is used to calculate the different GHG removal percentiles (see box 7.4):

1. Determine the size of the geographic area.

2. Sort each baseline candidate by its GHG removals in each time period (e.g., one year) from lowest to highest. For emission reduction project activities, the lowest-to-highest sorting order reflects the best-to-worst GHG performing units, and for LULUCF project activities, it reflects the worst-to-best performing land areas.

3. Label each land unit so that x_1 has the lowest GHG removals and x_a has the highest GHG removals, where
   - x_m is the GHG removals assigned to each land unit m, representing each baseline candidate.

   - a is the total land area represented in the geographic area.

   - m is the rank of each land unit with respect to its assigned GHG removals. Each land unit should have a distinct rank, and the rank is assigned sequentially to each land unit with the same GHG removals.

4. Determine the GHG removals corresponding to a specific percentile (pc) between 0 and 100 by
   - Calculating its approximate rank, w
   $w = (a \cdot pc)/100 + 0.5^1$

   - Assigning g to be the integer part of w and f the fraction part of w
   (e.g., if $w = 384.25$, then $g = 384$ and $f = 0.25$)

5. Calculate the GHG removals (pe) of the specific percentile (pc) using the following equation:
   $pe = (1-f) \, x_g + f \, x_{g+1}$

where

x_g is the GHG removal assigned to land unit g

For LULUCF project activities, the total land area, a, encompassed by the geographic area is usually large. Therefore, x_g and x_{g+1} likely correspond to the same type of baseline candidate. This means that the GHG removals for any given percentile probably correspond to the GHG removals of a specific type of baseline candidate.

These stringency calculations are repeated for each time period.

37

BOX 7.4 Sample Calculations of Specific Percentiles for a LULUCF Project

Each of four different types of baseline candidates represents a different land use, and each land use has the following GHG removals for each time period:

TYPE OF BASELINE CANDIDATES	A		B		C		D	
Land area represented by baseline candidates with the same carbon stocks (ha)	400		700		300		700	
	Carbon stock (t C/ha)	GHG removals (t CO_2/ha)	Carbon stock (t C/ha)	GHG removals (t CO_2/ha)	Carbon stock (t C/ha)	GHG removals (t CO_2/ha)	Carbon stock (t C/ha)	GHG removals (t CO_2/ha)
Time period 0	310		224		382		250	
Time period 1	308	-7.3	220	-14.7	384	7.3	256	22
Time period 2	302	-22	216	-14.7	388	14.7	267	33
Time period 3	297	-22	212	-14.7	392	14.7	288	77

The most stringent stringency level is represented by the highest GHG removals in each time period. Therefore, the most stringent stringency level is 22 t CO_2/ha in time period 1, 33 t CO_2/ha in time period 2, and 77 t CO_2/ha in time period 3.

The weighted mean GHG removals for time period 1 are calculated as

$$\frac{(-7.33 \cdot 400) + (-14.67 \cdot 700) + (7.33 \cdot 300) + (22 \cdot 700)}{(400 + 700 + 300 + 700)}$$

$$= 2.094 \text{ t } CO_2 \text{ /ha}$$

The percentile GHG removals are calculated as follows:
- Sort each hectare by its GHG removals, from lowest to highest and assign it a rank.

RANK, m, FOR EACH ha		1–700	701–1101	1102–1402	1403–2103
Assigned GHG removal, x_m (t CO_2/ha)	Time period 1	-14.67	-7.33	7.33	22
RANK, m, FOR EACH ha		1–400	401–1101	1102–1402	1403–2103
	Time period 2	-22	-14.67	14.67	33
	Time period 3	-22	-14.67	14.67	77

- To determine the 90th percentile GHG removals in time period 1:
  $$w = (2100 \cdot 90)/100 + 0.5 = 1890.50$$
  Therefore, $g = 1890$ and $f = 0.50$
  $$pe = (1-0.5) \cdot 22 + 0.5 \cdot 22 = 22 \text{ t } CO_2/\text{ha}$$

- To determine the 50th percentile (or median) GHG removals in time period 1:
  $$w = (2100 \cdot 50)/100 + 0.5 = 1050.50$$
  Therefore, $g = 1050$ and $f = 0.50$
  $$pe = (1-0.5) \cdot -7.33 + 0.5 \cdot -7.33 = -7.33 \text{ t } CO_2/\text{ha}^2$$

7.1.4 SELECT AN APPROPRIATE STRINGENCY LEVEL

An appropriate stringency level typically results in a performance standard that has higher than (weighted) mean GHG removals. The aim of the stringency level is to make sure that only those project activities representing better GHG management practices or higher GHG-performing land uses generate GHG reductions. Although there is no hard or fast rule for choosing a stringency level, there are a number of factors to consider when selecting it:

1. Regulatory requirements; for example, if legal drivers result in higher GHG removals, then a higher stringency level may be appropriate.

2. Recent and planned investments; for example, if there is a trend in the geographic area for land to be converted from agricultural uses to forest uses, then a higher stringency level may be appropriate.

CHAPTER 7: Estimating the Baseline GHG Removals—Performance Standard Procedure

3. Management regimes; for example, if there is a trend in forest management practices that lengthens harvest cycles in the geographic area (resulting in higher carbon stocks), then a higher stringency level may be appropriate.

Other factors to consider that are specific to reforestation and forest management project activities are the following:

Reforestation Project Activities

Higher stringency levels may be appropriate when agricultural management is turning toward higher sequestering management practices. For example, if no-till practices instead of conventional tillage practices are becoming more prevalent during the temporal range, a higher stringency would reflect the future trend by recognizing that there is increased carbon stored as a result of this management change. When the geographic area is predominantly an area of high productivity for wood production, then a higher stringency level may be more appropriate. For reforestation project activities in areas where wood production and the internal rate of return on the GHG project are low, then a lower stringency level may be appropriate, as planting may be a relatively unattractive land-use option.

Forest Management Project Activities

The stringency level for forest management projects represents a baseline management practice for the defined geographic area, which project developers use in conjunction with the project site's current inventory to define the baseline GHG removals. All aspects of the baseline management practice chosen for the performance standard should be described very clearly to ensure that the baseline GHG removals calculated on different project sites in the region will have consistent parameters, even if the starting inventory on the lands are different (see box 7.3).

A stringent management practice should reflect best practices in a region and may be most appropriate where there are numerous potential project sites and an active forestry industry, and/or forest management regulations are otherwise becoming more stringent. A less stringent management practice may be appropriate where there are fewer potential project sites and lower expectations on those sites in terms of productivity and regulations.

For example, if the performance standard is 22 t CO_2 removed/ha in time period 1 and the project site has 250 ha, the total baseline GHG removals in time period 1 are 22 t CO_2/ha x 250 ha = 5500 t CO_2 (see equation in box 6.1). Where appropriate, apply the land-use trend factor to the baseline GHG removals in each time period (see chapter 8). The GHG removals are then used in chapter 10, Monitoring and Quantifying GHG Reductions, to calculate the GHG reduction.

In addition, total baseline carbon stocks should also be calculated. Total baseline carbon stocks for forest management project activities are found by calculating the carbon stocks per unit of land area at the beginning of the project time period and adding any baseline changes in carbon stocks per unit of land area expected during the length of the GHG project. This total per unit of land area is then multiplied by the area covered by the project activity (see the equation in box 6.1).

For reforestation project activities where the performance standard represents only one type of baseline candidate for the project lifetime, the equation in box 6.1 may be used. For performance standards derived from different types of baseline candidates (i.e., land use) over time (as in box 7.4) finding the total carbon stocks does not make sense and need not be done. If the weighted mean GHG removals are used to define the performance standard, then the total baseline carbon stocks can be found by

- calculating the percent of the geographic area that each type of baseline candidate represents,

- applying these percentages to the area of the project site,

- multiplying the baseline carbon stocks by their respective area of the project site calculated above (see the Nipawin Afforestation Project for a full example of this calculation).

Guidance on quantifying carbon stocks and resources to use can be found in chapter 9, Estimating and Quantifying Carbon Stocks. The Nipawin Afforestation Project example also illustrates how to calculate changes in carbon stock, GHG removals and total carbon stocks.

7.1.5 ESTIMATE THE BASELINE GHG REMOVALS AND TOTAL BASELINE CARBON STOCKS

Baseline GHG removals are those GHG removals per unit area of land in each time period that correspond to the performance standard. The total baseline GHG removals are the GHG removals for the entire area of the project site in each time period.

NOTES

[1] This .5 is added to make sure that when the number of baseline candidates is uneven, the median baseline candidate will be represented by the baseline candidate in the middle of the distribution. For example, if there are seven baseline candidates, the median will be the fourth baseline candidate, not the third.

[2] Clearly, this stringency level would not be used, as it represents an emission of CO_2.

8 Applying a Land-Use or Management Trend Factor

The *land-use or management trend factor* (land-use trend factor) is an estimate of the rate at which land-use or management changes are occurring in the geographic area and during the temporal range defined in chapter 5, Identifying Baseline Candidates. It can be applied to both the performance standard and the project-specific baseline procedures to ensure that the baseline GHG emissions and removals more closely reflect an area's changing conditions.

8.1 When to Apply the Land-Use or Management Trend Factor

Two conditions are important to determining whether it is appropriate or possible to apply a land-use trend factor.

First, there should be at least two time periods, preferably at the beginning and the end of the temporal range, with reliable and specific data on lands in the geographic area (see box 8.1). Without reliable data from at least two time periods covering the geographic area, it is not possible to properly define the baseline candidates with the specificity required to estimate a land-use trend reliably or even to identify with certainty that there is a clear pattern of change. The more data that can be collected on the activities occurring in the geographic area, the greater the value will be of the derived land-use trend factor.

If there are no (or negligible) relevant land-use or management changes occurring in an area or there are insufficient data to estimate the land-use trend, the information then used to make these determinations should be clearly and transparently documented.

Second, the project activity and baseline candidates should represent discrete land-use or management changes and not be a variation in the level of intensity of the land use or implementation of a management practice. For example, in forest management project activities, various levels of forest thinning or differing lengths for harvest rotations within a geographic area would represent a change in the intensity of a management practice. Applying the land-use trend factor in this case therefore would not be appropriate because it would be difficult to discern the changes against which a rate could be applied.

BOX 8.1 Finding Relevant Land-Use and Management Trend Data

To ensure that the data being used to establish a land-use trend factor are relevant, consider the following:

1. Is there enough information to identify the specific units of land in the geographic area being used that have changed and what they have changed to? For example, a project developer may have found abundant information about baseline candidates based on census data or other surveys, but without specific information about the exact activities occurring and which lands have been affected, the project developer may have difficulty ascertaining whether the trend is relevant to the project activity or how the information should be applied to adjust the baseline GHG removals.

2. Is there enough information about the baseline candidates to distinguish them sufficiently when discerning trends? For example, there may be several "afforestation" activities occurring in a region, but is enough known about these activities to determine why they have been occurring and whether there are any differences among them?

3. Is there enough information to find out why units of land and the underlying specific management practices or activities in the geographic area are shifting? The project developer needs to ascertain that the temporal range being used to collect data on the trends either includes or excludes (depending on current conditions) laws, regulations, or other programs that may be causing the changes. Although this information should have been found in chapter 5, it must be considered again in this chapter, especially as it may relate to land units that have changed because of climate change programs or incentives.

GHG programs run by local, regional, or national governments are in the best position to gather the data required to develop a land-use trend factor. Specifically, a GHG program would be in the best position to decide whether land units converted as a result of its or other government activities should be included or excluded from the analysis of the trend.

8.2 Estimating the Land-Use or Management Trend Factor

8.2.1 LAND-USE OR MANAGEMENT TREND METRIC

The appropriate metric to use when estimating the land-use trend factor is the percent change in land area to a given land use or management practice for each time period. Specifically, for reforestation project activities, it is the percentage of land converted from the baseline candidate's land use(s) to the project activity's land use for each time period. For forest management project activities, it is the percentage of land converted from the specified management practice(s) to the project activity's management practice for each time period. In some cases, baseline candidates may be shifting to multiple land uses or practices. It is important to track those baseline candidates shifting to a land use or practice similar to the project activity, but other shifts may be tracked as well.

8.2.2 ESTIMATING THE LAND-USE OR MANAGEMENT TREND FACTOR

One of two main approaches can be used to estimate a land-use or management trend: a simple approach that extrapolates past land-use or management change data and a more complicated approach that uses the drivers of change as well as past land-use or management data to develop a trend. A variety of methods and models[1] can be used for either of these approaches, including statistical and trend analysis, modeling, default values, and remote sensing.

Extrapolating Past Land-Use or Management Trends Using Historical Data

This approach uses historical data to determine the trend of land-use or management change. The trend is estimated using either tendency or policy analysis and historical land-use or management change data within the geographic area to project the trend into the future. The results depend on the temporal range and geographic area chosen (and/or data availability) for the analysis. Although this is a relatively straightforward approach, it cannot detect any of the underlying causes or drivers of land-use or management change that may influence some land-use patterns or changes in management practices.

Extrapolating Past Land-Use or Management Trends Using Drivers of Land-Use or Management Change

Modeling drivers of land-use change either can extrapolate past trends using socioeconomic drivers or, using geographic information systems (GIS), can extrapolate past trends using spatially based biophysical and socioeconomic drivers. Some of the common drivers of land-use and management change are:

- Physical factors, for example, elevation, slope, aspect, soil type, rivers, navigable water and watersheds.

- Development factors, for example, major and minor roads, dam construction.

- Urban and rural population.

- Political boundaries.

- Centers of commerce, logging camps, and communities.

- Land tenure and current land-use distribution.

Many models can be used to determine the relative importance of the drivers influencing land-use or management change, although currently no one model is superior in all circumstances. Some desirable components of a model using drivers to estimate rates of change are:

- Transparency.

- A routine for determining the important land-use or management drivers in a given area.

- Empirical calibration.

- Internal checks such as calibration of the fit.

- Validation routine.

- Reliance on generally available data (e.g., government statistics and maps).

- Ability to generate probability (or likelihood) of change maps for the area.[2]

8.3 Applying the Land-Use or Management Trend Factor

The land-use trend factor is applied to adjust the baseline GHG removals in each time period found by identifying either the performance standard or the baseline scenario. Besides understanding the land-use or management trend, the project developer also needs to understand the GHG removal or emission impacts resulting from the trend. Some of this information may already have been collected, especially if the performance standard procedure was used. But project developers should be prepared to defend not only the trend factor identified but also the GHG emissions or removals added to the baseline GHG removals as a result of its application.

NOTES

[1] Some of the many models available to estimate the rate of land-use or management change are the FAC model (FAO 1993; Scotti 2000), LUCS (Faeth et al. 1994), and GEOMOD (Hall et al. 1995, 2000). Use the model that best suits the GHG project's conditions and data availability.

[2] Personal communication from Sandra Brown, Winrock International, and Ben de Jon, El Colegio de la Frontera Sur.

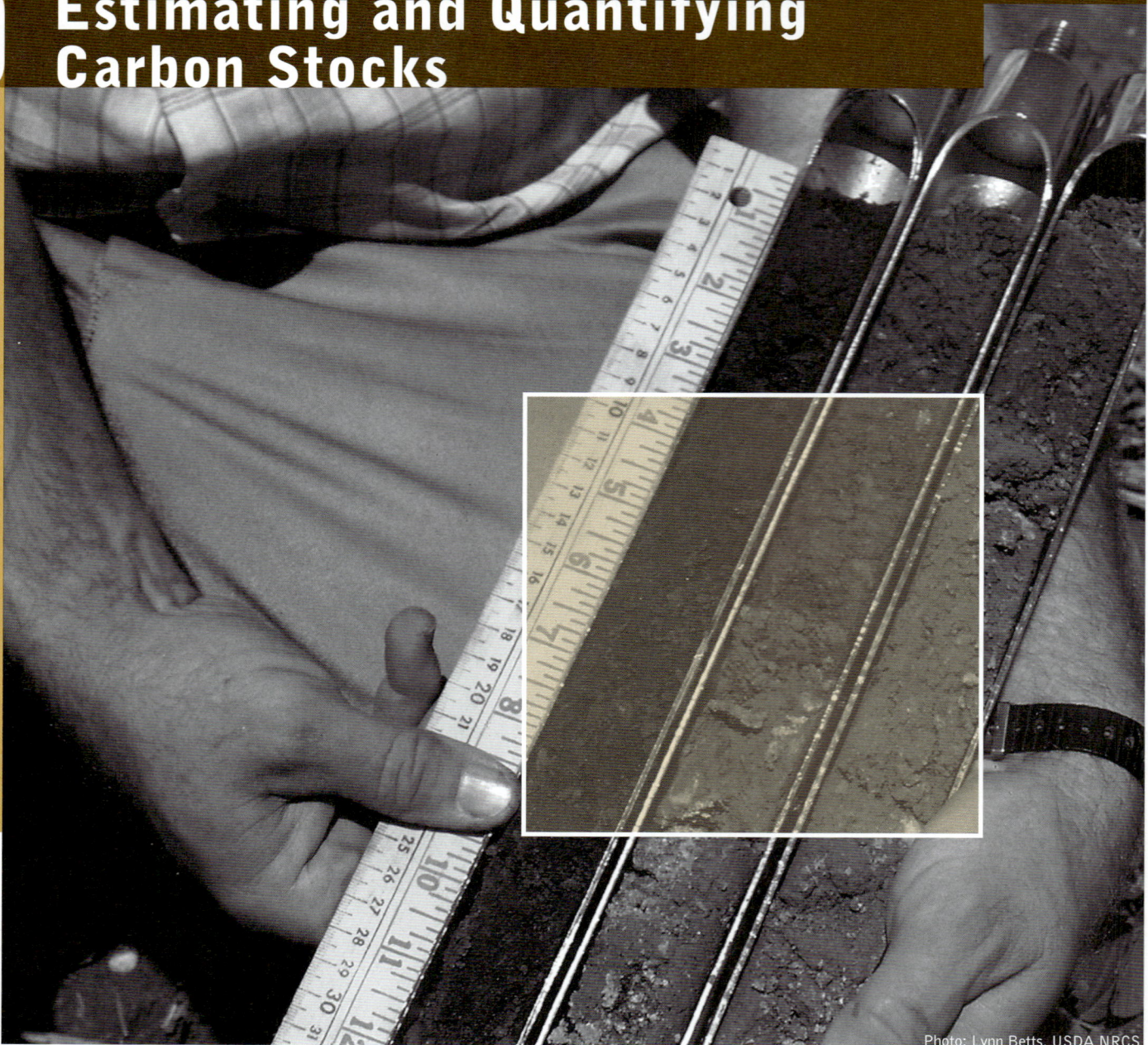

Photo: Lynn Betts, USDA NRCS

9 Estimating and Quantifying Carbon Stocks

This chapter provides a conceptual overview of estimating and quantifying the baseline and project activity carbon stocks used to find both the change in carbon stocks needed to quantify GHG removals and the total carbon stocks. It describes both the calculations for an ex-ante estimation for a project's potential carbon stocks and an ex-post quantification for its verification. This is not meant to be a comprehensive discussion but, rather, an outline of the important aspects to consider when estimating or quantifying carbon stocks and some resources that provide the required detail to make such measurements or estimations.

44

Some of the important considerations in estimating or quantifying carbon stocks are:
- Identifying the carbon pools to measure or estimate.

- Estimating or measuring the carbon stocks, including the carbon stocks at the beginning of the project activity's implementation and any carbon stock changes expected during the project lifetime. If measuring carbon stocks, then determining the type, number, location of sample plots used to measure carbon stocks, and the frequency with which to measure the plots is also important.

- Estimating and adjusting for uncertainty.

9.1 Identifying the Carbon Pools to Measure or Estimate

Carbon is stored in several components of a biological ecosystem. Each is referred to as a carbon pool, and together they comprise the carbon stocks of a biological system like a forest stand.

The common carbon pools to consider are:
- Living biomass: aboveground biomass and belowground biomass (e.g., stems, branches, foliage, and roots).

- Dead organic matter: dead wood and litter.

- Soils.

Table 9.1 lists some recommended carbon pools to estimate or quantify and monitor for LULUCF project activities. The IPCC Good Practice Guidance (2003) and the LULUCF Sourcebook (Pearson, Walker, and Brown 2005) provide more details about these carbon pools.

TABLE 9.1 Decision Matrix of the Main Carbon Pools to Estimate or Quantify and Monitor for LULUCF Project Activities

PROJECT TYPE	MAJOR CARBON POOLS						Wood Products
	Live biomass			Dead biomass			
	Trees	Herbaceous	Roots	Fine	Coarse	Soil	
Avoid emissions							
Stop deforestation	Y	M	R	M	Y	R	M
Reduce impact logging	Y	M	R	M	Y	M	N
Improve forest management	Y	M	R	M	Y	M	Y
Sequester carbon							
Plantations/reforestation	Y	N	R	M	M	R	Y
Agroforestry	Y	Y	M	N	N	R	M
Soil carbon management	N	N	M	M	N	Y	N
Carbon substitution							
Short-rotation energy plantations	Y	N	M	N	N	Y	A

Source: Adapted from Brown 1999; Brown, Masera, and Sathaye 2000.

Notes: A = Stores carbon in unburned fossil fuels.

Y = yes and indicates that the change in this pool is likely to be large and should be measured.

R = recommended and indicates that the change in the pool could be significant, but measuring costs to achieve desired levels of precision could be high.

N = no and indicates that the change is likely to be small to none, and thus it is not necessary to measure this pool.

M = maybe and indicates that the change in this pool may need to be measured depending on the forest type and/or management intensity of the project.

9.2 Ex-Ante Estimation versus Ex-Post Quantification

An ex-ante estimation of carbon stocks is needed when a project developer is submitting project documentation to a GHG program or determining the magnitude of potential GHG reductions for project or financial-planning purposes. For the project activity, this is only an initial estimate of carbon stocks, whereas for the baseline carbon stocks it can be an estimate or be based on direct measurements.

An ex-post quantification of carbon stocks differs from ex-ante estimation in that carbon stocks are quantified according to actual measurements. Because of the importance of this determination, ex-post quantification is based, to the extent possible, measured carbon stocks. How the ex-post quantification is done should be clearly documented in the monitoring plan (see chapter 10, Monitoring and Quantifying the GHG Reductions).

When existing land-use or management practices on the project site are used to determine the baseline carbon stocks, the carbon stocks should be measured at least once. When the baseline carbon stocks are not based on existing land-use or management practices or the baseline carbon stocks are dynamic, control plots that represent the baseline conditions may be established, and the carbon stocks on these sites are measured periodically.

For LULUCF project activities, annual measurements are not practical because of the relatively slow rate of carbon change in vegetation and soil. Instead, measurement every three to five years may be sufficient. The frequency of measurement may also depend on the GHG program's rules and ecosystem characteristics. Measurements are necessary for the lifetime of the project,[1] which is likely to be several decades, so the measurements form the basis of a long-term monitoring program. Quantification of the GHG removals however may still occur annually even if not measured annually.

9.3 Methods of Estimating or Quantifying Carbon Stocks

Methods of estimating or quantifying carbon stocks include direct measurement, values taken from the scientific literature, carbon models, or some combination of these. They can be used for both ex-ante estimation or ex-post quantification but are applied slightly differently in each case. Regardless of the protocol or guidance used to measure carbon stocks, the transparency of the method(s) followed and the subsequent results are important.

9.3.1 DIRECT MEASUREMENT OR SAMPLING

When carbon stocks are being estimated only ex-ante, for example, for baseline carbon stock calculations, direct measurements may not be needed. The exception may be when no carbon stock data exist for the geographic area, in which case some measurements may be necessary. If ex-ante measurements are being taken for the project activity, the measurements need not be as accurate and precise as those for ex-post quantification, which are the measurements that must be verified. Box 9.1 describes accuracy, precision, and conservativeness in direct measurement or sampling.

With ex-post quantification all carbon pools need to be quantified unless the project developer can show that a pool will not be a source of GHG emissions or will not change during the life of the project (for an indication of which pools should be considered, see table 9.1).

> **BOX 9.1 Accuracy versus Precision and Conservativeness**
>
> It is important that data collected through direct measurements or sampling are accurate and/or precise.[2]
>
> Accuracy is one of the principles for GHG accounting (see chapter 2, Key Concepts and Principles for LULUCF Projects). For direct measurements or sampling, accuracy is defined as the proximity of the sample measurements to the actual value. Therefore, the level of accuracy is how close the true value is to any repeated measurements or estimates of the carbon stocks.
>
> Precision refers to how well a value is defined. For direct measurements and sampling, precision is how close the repeated measurements or estimations are to the same quantity of carbon stocks. This is represented by how closely the results from the various sampling points or plots are grouped. But a precise measurement may also be inaccurate, for example, when the sampling equipment or the sampling design is systematically biased. Measurements of carbon stocks should be both accurate and precise so as to inspire the confidence in the results.
>
> An important principle linked to estimating or quantifying carbon stocks is conservativeness. Sometimes a particular carbon pool cannot be measured or must be estimated, in which case a conservative estimate of the carbon stocks associated with that carbon pool should be used. For example, if only an inaccurate measurement of the belowground living biomass for the project activity is possible, then being conservative means reporting the lower bound of the confidence interval. In contrast, being conservative for the baseline carbon stocks means using the higher bound of the confidence interval. The reported carbon stocks then are lower than the mean. Understanding the confidence intervals around various measurements and/or calculations helps estimate the overall uncertainty associated with quantifying the GHG reduction.

BOX 9.2 **The Role of Direct Measurement in Determining GHG Credits**

Ex-post quantification should be based as much as possible on measured values. To the extent that project developers can measure GHG removals to submit to the GHG program, these values often will need to be verified by a third party. The verification is usually the basis for awarding credits to the project. Verification includes a review of the accounting protocol used (e.g., the identification of the baseline scenario), quantification protocol (e.g., the quantification of the baseline and project activity GHG removals), monitoring plan (e.g., monitoring reversals of carbon storage), and quality assurance and quality control (QA/QC) procedures (e.g., data collection and storage procedures, carbon sampling procedures). Project developers will need to keep detailed accounts of all their accounting and quantification protocols, measurements, etc. Chapter 12 provides a detailed list of the information required to verify the methodologies for the accounting, quantifying and monitoring of the GHG reductions. Annex C provides a detailed list of information needed to ensure QA/QC procedures can be verified.

Ex-post quantification is typically carried out by the project developer and may need to be verified by an independent third-party auditor (for a discussion of the role of direct measurement in determining GHG credits, see box 9.2). Therefore, a rigorous quantification protocol should be used to ensure accurate and consistent results.[3] Decisions about the type of plots used to measure carbon stocks, the number and location of the plots, and the frequency with which to measure the plots all are important when quantifying carbon stocks. Detailed guidance on establishing sampling protocols for carbon stock determinations is given in the IPCC Good Practice Guidance for LULUCF 2003, MacDicken 1997, and Pearson, Walker, and Brown 2005. Other protocols or regional guidance on sampling protocols may be used as well.

For those LULUCF project activities in which baseline carbon stocks are expected to change substantially during the project's life and a baseline scenario has been identified using the project-specific procedure, control plots can also be established. For example, when the baseline scenario is already taking place within the geographic area, control plots can be monitored over the project's life in order to capture any changes in the baseline conditions. Control plots are less useful for the performance standard procedure, as enough control plots must be established so that each type of baseline candidate identified is represented.

9.3.2 DEFAULT VALUES

The availability and applicability of default values vary widely depending on the vegetation species, soil types, and geographic region. In some cases, extensive and detailed data are readily available, for example, soil carbon data from soil surveys in the North American Great Plains region. In other cases, though, data have not been collected or are not easily accessible, so those from previous research or surveys are generally acceptable for estimating carbon stocks.

Because direct measurement usually is expensive, project developers should try to find existing sources of data for ex-ante estimates. Even though ex-post quantification emphasizes measured values, in some instances default values also may be acceptable. The IPCC Good Practice Guidance for LULUCF (2003) provides guidance on when default values may be used for ex-post quantification, additional sources of information are available in the references section of this document.

Reforestation Project Activities

Reforestation is typically carried out on agricultural land, with soil carbon being one of the carbon pools that should be estimated or measured. For some regions, soil survey data are available, are generally of high quality, and often include measurements of soil carbon, which can be used to estimate the baseline soil carbon stocks and carbon stock changes.

Forest Management Project Activities

Carbon stocks may be estimated using predefined growth or yield curves for a certain area and tree type/site class when direct measurements are not possible or are too costly. These growth or yield curves are typically based on a detailed analysis of the growth patterns of specific species in a given area and may be useful to estimate the carbon stocks from each baseline candidate. Other carbon pools, such as the root and dead organic matter pools, can be estimated using equations based on the volume of biomass and information about the structure and form of trees, climate, and the like, instead of using measured values.

9.3.3 MODELS

The number of models to be used for estimating carbon stocks has grown rapidly. Simpler models like CO_2FIX (Masera et al. 2003, http://www2.efi.fi/projects/casfor/) and the Carbon On-Line Estimator (which uses U.S. Forest Service FIA data), have relatively modest data requirements, whereas more complex ecosystem models, like BIOME-BGC (Thornton et al. 2002) or CENTURY (Parton et al. 1987) require extensive data on species physiology and carbon profile.

47

9.4   Accounting for Uncertainty

Uncertainty should be considered for the following LULUCF project activities:

- Estimation of the baseline GHG removals and secondary effects.

- Estimation of the project activity's GHG removals and secondary effects.

- Risk of carbon stocks being reversed through anthropogenic or natural events. This is discussed in chapter 11, Carbon Reversibility Management Plan.

When estimating or measuring baseline or project activity GHG removals and secondary effects, uncertainty can arise from

- Scientific uncertainty, caused by insufficient knowledge of the processes sequestering carbon or emitting GHGs.

- Estimation uncertainty, caused by either (1) model uncertainty, when there is uncertainty around the mathematical equations used to characterize the relationships among various processes or variables, or (2) parameter uncertainty, when there is uncertainty around the quantification of the parameters used as inputs (e.g., activity data, emission factors, field measurements).

When possible, the project developer should assess both estimation and parameter uncertainty related to the GHG project. The two types of parameter uncertainty are systematic uncertainty and statistical uncertainty.

Systematic parameter uncertainty results when the data are systematically biased. In other words, the average measured or estimated value is always greater than or less than the true value. For example, a bias may result if the emission factors are developed from unrepresentative samples, not all the relevant emission sources or sinks have been identified, or incorrect or incomplete estimation methods or faulty measurement equipment has been used.

Sources of systematic parameter uncertainty should always be identified and reported qualitatively. If possible, the direction of bias and its magnitude should be documented, as well as any mitigation options to reduce the uncertainty.

Statistical parameter uncertainty is random and pertains to discrepancies in the data used to quantify the GHG reductions. It results from human error, fluctuation in the measurement equipment, and so forth.

Random uncertainty can be detected by discrepancies in repeated samples or measurements of the same carbon pool. Ideally, random uncertainties should be statistically estimated using available empirical data, for example, from repeated sampling. But if not enough sample data are available to develop valid statistics, statistical parameter uncertainties can be estimated using methods like those described in IPCC 2001 or the GHG Protocol Uncertainty Tool.

Models can provide reasonable estimates of carbon stocks if

- The model is accepted in the scientific community.

- The appropriate input data are available and have been used, for example, soils, climate, and vegetation growth.

- The model's parameter values reflect local conditions.

- The model's results have been tested against independent data representing local conditions.

Data for validation and calibration may be difficult to obtain, depending on the region and also on the model's sophistication. If the required data are available, models may provide reasonable values for carbon stocks. But remember that model results, at best, are only estimates and cannot substitute for real data.

One advantage of models is that the carbon stock changes and GHG removals often can be automatically calculated from the carbon stock data generated by the model.

Many tools and guidance documents have been developed to estimate the uncertainty levels associated with carbon stocks. These tools and documents typically also describe mitigation options to reduce the level of uncertainty around carbon stock estimates or measurements. Some tools and guidance resources to use are:

- MacDicken (1997), A Guide to Monitoring Carbon Storage in Forestry and Agroforestry Projects.

- IPCC (2001), Good Practice Guidance and Uncertainty Management in National Greenhouse Gas Inventories.

- IPCC (2003), section 5.2 of Good Practice Guidance for Land-Use, Land-Use Change and Forestry.

- Pearson, Walker, and Brown (2005), Sourcebook for Land-Use, Land-Use Change and Forestry Projects.

- WRI/WBCSD Uncertainty Tool for Corporate Inventories (www.ghgprotocol.org).

Although some of these resources are not specifically for project-level accounting, they still can be used to assess a project's uncertainty levels.

When a measurement, calculation, or estimation continues to produce a high level of uncertainty, the project developer should apply the conservativeness principle and use the most conservative value within the uncertainty range or confidence interval being considered. This is especially important when estimating the baseline carbon stocks, for which direct measurement and/or repeated sampling to define accuracy and precision may not be possible. Because project activity carbon stocks are monitored more frequently, the uncertainty around these measurements should be smaller.

NOTES

[1] The life of a project is generally set by the GHG program's liability period. For example, the GHG program may set its liability period for LULUCF projects as twenty years, which also will be the length of the project life.

[2] For more information about accuracy, precision, and conservativeness, see Pearson, Walker, and Brown 2005.

[3] GHG programs usually have specific quantification rules for each project type.

Photo: Tim McCabe, USDA NRCS

10 Monitoring and Quantifying GHG Reductions

Monitoring ensures that the estimated emission reductions or carbon enhancements are actually taking place as projected and that the baseline assumptions underlying the estimations remain relevant over the project's lifetime. A monitoring plan is a working document that describes the procedures for collecting data related to the project activity and baseline parameters and controls the quality of the collected data.

10.1 Creating a Monitoring Plan

Many resources can be used to help devise a monitoring plan, for which IPCC 2003, the IPCC Good Practice Guidelines 2006, MacDicken 1997, and Pearson, Walker, and Brown 2005 are commonly cited.

The monitoring plan should be complete, consistent, and transparent. It should describe the steps that the project developer will take to

- Monitor the project activity's primary and secondary effects.

- Monitor the baseline parameters.

- Maintain quality assurance / quality control (QA/QC) practices.

When drawing up a monitoring plan, the project developers must make trade-offs between the accuracy and the associated costs of rigorous monitoring. When deciding which carbon stocks, GHG emissions, and baseline parameters to measure and how to measure them, project developers should follow the principles outlined in chapter 2, Key LULUCF Accounting Concepts and Principles, particularly the principle of conservativeness. For additional guidance on setting an appropriate level of accuracy, see Pearson, Walker, and Brown 2005, section 6, Developing a Measurement Plan.

10.1.1 MONITORING PROJECT ACTIVITY PRIMARY AND SECONDARY EFFECTS

In addition to the information in the Project Protocol about quantifying a project activity's primary and secondary effects, the monitoring plan for LULUCF project activities should include the following:

- Which secondary effects will be measured (for guidance on identifying significant secondary effects, see chapter 3, Defining the GHG Assessment Boundary).

- Which sampling sites and carbon pools will be measured (for guidance on and resources for selecting sampling sites and carbon pools, see chapter 9, Estimating and Quantifying Carbon Stocks).

- The methods used to measure carbon pools (e.g., estimation, direct measurement, models).

- The frequency with which the carbon stocks in each carbon pool will be measured or modeled (e.g., carbon stocks are modeled each year, with direct measurements every ten years), indicating those carbon pools that should be monitored more often.

- The level of acceptable uncertainty in carbon stock measurements for sampling sites and how this uncertainty will be accounted for in the final determination of GHG removals and total carbon stocks.

- The emission factors or default factors used and their sources.

- A description of the equipment used to gather data, the control sites (if any), and the procedures used to maintain their accuracy/relevance.

- Any unintentional releases of stored carbon stocks, for example, losses due to fire, pest infestation, crop/tree disease, and illegal logging (for guidance on accounting for the loss of carbon stocks, see chapter 11, Carbon Reversibility Management Plan).

- Insignificant secondary effects. Although some secondary effects may seem to be too small to warrant monitoring, they still should be monitored (though much less rigorously) to ensure that their effect remains insignificant throughout the life of the GHG project.

If secondary effects or carbon pools are omitted because of cost or any other constraint, the monitoring plan should explain these exclusions.

10.1.2 MONITORING BASELINE PARAMETERS

The monitoring plan should include any parameters used to estimate baseline GHG removals and emissions (e.g., emission factors) and the assumptions used to determine the baseline GHG removals and emissions (e.g., land use identified as the baseline scenario for reforestation project activities). The list of parameters that project developers should consider is:

- Any changes in the initial emission factors or default factors used to estimate the baseline carbon stocks changes or secondary effects.

- If control plots were used, the characteristics of the control plots over time. This ensures that the plots still represent an accurate proxy for the baseline scenario.

- The land-use or management change trend in the geographic area. The monitoring plan should include provisions for monitoring the trends in those land uses, management practices, or markets used to determine the baseline GHG removals.

- Outbreaks of insects, fire, and disease. When these events occur, the baseline GHG removals should be adjusted accordingly.

10.1.3 DESCRIBING QA/QC MEASURES

The monitoring plan should encompass the QA/QC procedures that were implemented to ensure that the data collection and calculations are accurate and complete. At a minimum, the plan should include the information listed in the Project Protocol's chapter 10, section 10.1.3, Describing QA/QC Measures. Annex C in this document offers a more detailed list of the information to include in a LULUCF QA/QC report.

10.2   Quantifying GHG Reductions

GHG reductions are calculated as the difference between project activity GHG removals and the baseline GHG removals for a given unit of land area (e.g., hectare) and specified time period. Throughout this document, GHG removals are found in terms of carbon dioxide, so when finding the GHG reduction, all carbon calculations must be translated into carbon dioxide equivalents (CO_2 eq).

Although the Project Protocol outlines GHG reductions in terms of carbon stocks, it is actually the change in carbon stocks that is used to estimate the GHG reductions from a LULUCF project. Therefore, the following equations are based on GHG removals, not carbon stocks.

The variables used to quantify GHG reductions are:

k: carbon pool

p: primary effect

z: project activity

s: secondary effect

t: time period

GHG Reduction

$$\text{GHG Reduction}_t \text{ (t } CO_2 \text{eq)} = \sum_z \text{Project Activity Reduction}_{zt}$$

where

$$\text{Project Activity Reduction}_{zt} = \text{Primary Effect}_{zt} + \text{Secondary Effect}_{zt}*$$

*Secondary effects can be GHG emissions or GHG removals. If they are GHG emissions, they should be written as a negative number, so they are subtracted from the primary effects removals.

Primary Effect

For LULUCF projects, the primary effect includes only the storage and releases of carbon through biological process due to land use and/or management practices.

$$\text{Primary Effect}_{zt} \text{ (t } CO_2 \text{eq)} = \text{GHG removals}_{zt}$$

$$\text{GHG Removals}_{zt} \text{ (t } CO_2 \text{eq)} = \sum_p [\text{Project Activity GHG Removals}_{pzt} - \text{Baseline GHG Removals}_{pzt}]$$

> Project Activity GHG Removals$_{pzt}$
> $= (\text{Project Activity Carbon Stocks}_{pzt} - \text{Project Activity Carbon Stocks}_{pz(t-1)}) \cdot \frac{44}{12} \text{t } CO_2/\text{t carbon}$

>> Project Activity Carbon Stocks$_{pzt} = \sum_k$ carbon stocks from each biological carbon pool measured, k, related to each primary effect, p, for project activity, z, in period t for a given unit of land area (e.g., hectare)

> Baseline GHG Removals$_{pzt}$[1] $= (\text{Baseline Carbon Stocks}_{pzt} - \text{Baseline Carbon Stocks}_{pz(t-1)}) \cdot \frac{44}{12} \text{t } CO_2/\text{t carbon}$

>> Baseline Carbon Stocks$_{pzt} = \sum_k$ baseline carbon stocks from each biological carbon pool measured, k, related to each primary effect, p, for project activity, z, in period t for a given unit of land area (e.g., hectare)

Secondary Effects

Only GHG emissions will need to be calculated for secondary effects unless a market response provokes a change in carbon stored from biological processes off the project site.

Secondary Effects$_{zt}$ = Emissions Secondary Effects$_{zt}$ + Δ Secondary Effects Carbon Stocks$_{zt}$

Emissions Secondary Effects$_{zt}$ = $\sum_s$ [Baseline Emissions$_{szt}$ − Project Activity Emissions$_{szt}$]

> Baseline Emissions$_{szt}$ = Baseline GHG emissions related to each secondary effect, s, for each project activity, z, in period t for a given unit of land area (e.g., hectare) (in t CO_2eq)

> Project Activity Emissions$_{szt}$ = GHG emissions related to each secondary effect, s, for each project activity, z, in period t for a given unit of land area (e.g., hectare) (in t CO_2eq)

Δ Secondary Effects Carbon Stock$_{zt}$ (t CO_2eq) =
Net Δ Secondary Effect Carbon Stocks$_{zt}$ • $\frac{44}{12}$ t CO_2/t carbon

> Net Δ Secondary Effect Carbon Stocks$_{zt}$ (t carbon)
> = $\sum_s$ [Δ Project Activity Secondary Effect Carbon Stocks$_{szt}$ − Δ Baseline Secondary Effect Carbon Stocks$_{szt}$]

>> Δ Project Activity Secondary Effect Carbon Stocks$_{szt}$
>> = Project Activity Secondary Effect Carbon Stocks$_{szt}$ − Project Activity Secondary Effect Carbon Stocks$_{sz(t-1)}$

>> Project Activity Secondary Effect Carbon Stocks$_{szt}$ = $\sum_k$ carbon stocks from each biological carbon pool measured, k, related to each secondary effect, s, for project activity, z, for period t for a given unit of land area (e.g., hectare)

>> Δ Baseline Secondary Effect Carbon Stocks$_{szt}$
>> = Baseline Secondary Effect Carbon Stocks$_{szt}$ − Baseline Secondary Effect Carbon Stocks$_{sz(t-1)}$

>> Baseline Secondary Effect Carbon Stocks$_{szt}$ = $\sum_k$ baseline carbon stocks from each biological carbon pool measured, k, related to each secondary effect, s, for project activity, z, for period t for a given unit of land area (e.g., hectare)

NOTES

[1] Although baseline GHG removals should already have been calculated in chapter 6 or 7, the equation is repeated here for reference.

53

11 Carbon Reversibility Management Plan

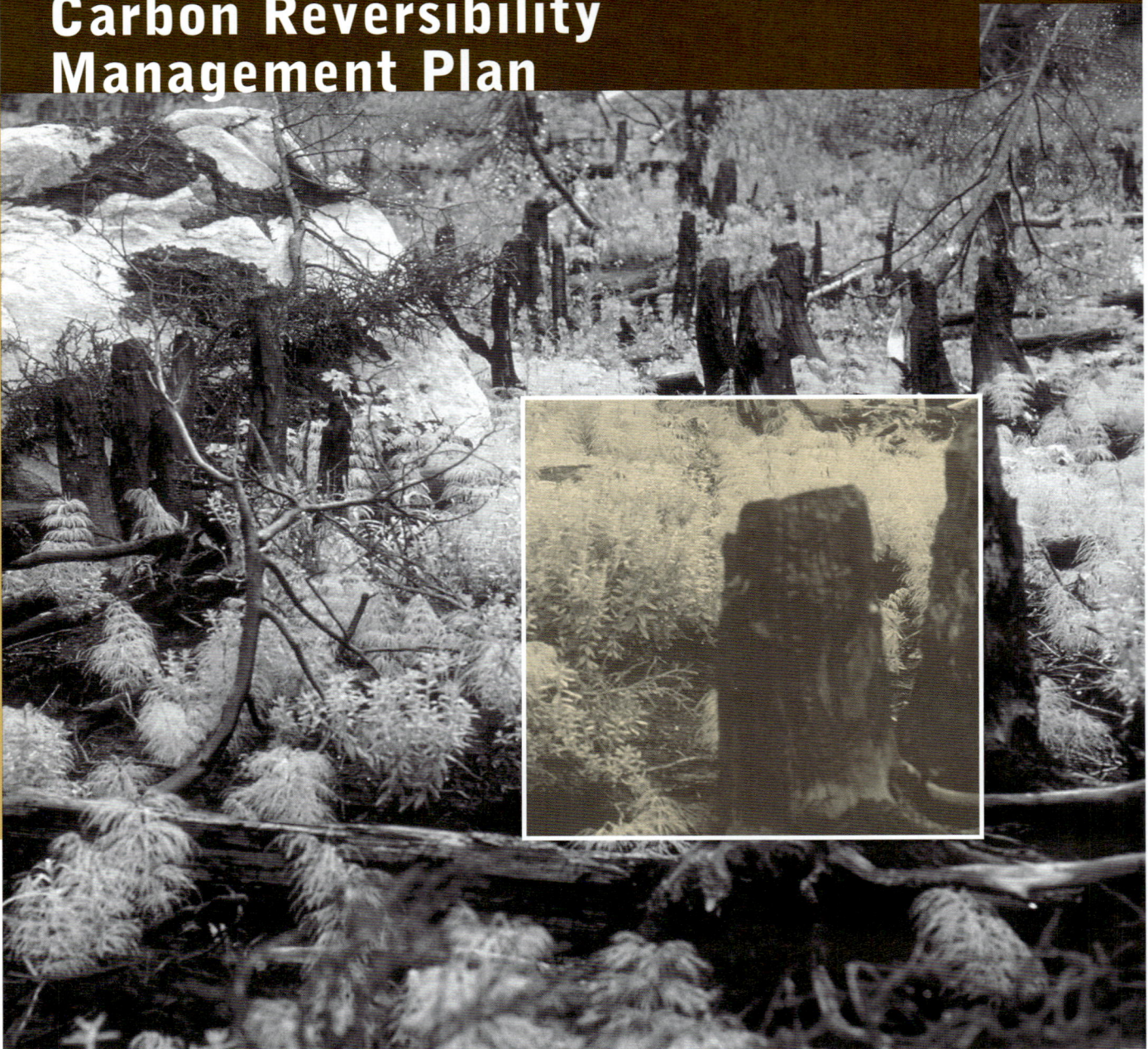

GHG reductions for LULUCF projects are unique in that the carbon dioxide levels reduced by removing and storing carbon in biological systems are temporary and the removed/stored carbon may return to the atmosphere in the future (referred to as a *reversal of carbon storage*). Reversals may result from intentional activities (e.g., planned harvesting of a forest) or accidents (e.g., forest fires, insect attacks). Intentional reversals should be factored into the assessment of the project's primary effect, as shown in chapter 10, Monitoring and Quantifying GHG Reductions, taking into account any compensating mechanisms incorporated into the project design.

The likelihood of a reversal varies according to project type, species, location, intensity of management, and availability of suppression or mitigation options. Project developers should consider the probability of reversals and how they plan to deal with them, and develop a *carbon reversibility management plan*. The plan should include methods for reducing the risk of reversals (mitigation) and replacing the carbon once it has been released (adaptation). To promote transparency and increase investors' confidence, any project that stores carbon in biological systems should include such a plan. It should document the reversible elements of the GHG project, assess the significance of the reversibility to the expected GHG reduction, and describe the measures that will be taken to reduce and/or compensate for the carbon reversal. This plan should be incorporated into the overall project documentation, as it will help GHG programs, investors, and others make more informed judgments about the project's effectiveness in reducing GHG emissions.

To prepare a carbon reversibility management plan,
- Identify and assess the reversible elements of the project's GHG reduction.

- Describe the actions undertaken to suppress or mitigate the reversibility of GHG reductions.

- For any residual risk of reversibility, establish mechanisms to compensate for any reversals of carbon storage.

- Develop a plan to monitor carbon reversibility and risk.

11.1   Identify and Assess the Reversible Elements of the Project's GHG Reduction

It is best to identify and assess the extent to which a project's GHG reduction can be reversed over time early in the project development process, as this will enable the project developer to take measures to reduce or eliminate reversibility during the project design phase. The first step is to determine which pools store carbon. In LULUCF project activities, carbon is typically stored in many different pools (e.g., living biomass, dead organic matter, soil), but only a subset of these pools may be at risk of reversal. For instance, soil carbon and belowground biomass (e.g., roots) may be minimally affected by a fire or even be enhanced by the addition of charcoal after a fire. Some GHG projects may also have multiple project activities and different land parcels, each with a different level of risk.

The next step is to consider the reversibility of each of the pools storing carbon (see box 11.1). At one extreme, GHG projects may be able to provide relatively secure, long-duration reductions, for example, a project in a government-designated wilderness area with high rainfall, low fire, and pest risk, secure land tenure enforced by the courts, and no harvesting. At the other extreme, the risk of

reversibility may be quite high due to variable outbreaks of fire and pests, rapid land-use change, insecure land tenure not enforced by the courts, fluctuating market and commodity prices and conditions, and uncertain and variable government policies. Most GHG projects fall somewhere in the middle, meaning that the risk of reversal should be addressed.

BOX 11.1 Reversibility and Baseline Carbon Stocks

Carbon storage may be reversed in the baseline carbon stocks and should be explored when estimating the baseline carbon stocks. For example, the existing data from forest inventories often indicate regional fire patterns and pest frequency, harvest patterns and methods, and biomass harvested per hectare. Regional patterns of resource and land use, socioeconomic trends, and current land-use practices may already have incorporated many of these reversibility risks into the performance benchmarks. Any intentional or planned activities affecting carbon stocks also should be incorporated into the calculation of the GHG reduction (e.g., thinning and harvesting for forest management projects).

Last, if the GHG reductions could be reversed, the risk should be estimated. To do this,
- Estimate the history of risk for the project activity, carbon pool, and geographic area over at least ten years, a period that will vary according to site and location.

- Determine the change in the stored carbon related to this risk (e.g., percentage of biomass burned per hectare for a particular forest type and location).

Risk estimates are typically quite uncertain, so they are usually presented as a range of probabilities. Project developers should use risk estimates that reduce the chance of underestimating the risk. The project's risk of reversal, a list of the project activities or carbon pools at risk of being reversed, the factors contributing to this risk, and the degree of uncertainty surrounding the estimate should all be documented. If the carbon storage is not reversible, this should be explained as well.

11.2 Describe Actions to Suppress or Mitigate the Reversibility of GHG Reductions

A number of actions can reduce the reversibility of carbon storage:

- **Easements.** Easements establish perpetual land-use or management restrictions on land deeds that bind future landowners to them. They may be relevant to some forest management and land-use change projects. When the ownership changes, easements can help avoid land-use conversion and/or changes in management that would result in releasing the stored carbon into the atmosphere in the future.

- **Project type, site selection, and prevention measures.** Such measures may be

  1. Choosing or planning GHG projects with a low risk of reversal. For example, one strategy for forest project planning is to take a landscape view of the project and to design plantation establishment in a staggered manner so that not all of the area reaches a harvestable age at the same time. In this way there will always be carbon stocks on the landscape even when some areas are being harvested. Therefore, a reforestation project thus designed will always have GHG removals and carbon stocks greater than those of the agricultural land use, so the carbon stock increase is "permanent," even with harvesting (except during catastrophic reversals such as fire).

  2. Selecting sites with a relatively low risk compared with that of other areas. In order to identify areas that have had fewer incidents, project developers must become familiar with the history of natural disturbance in their region.

  3. Adding prevention measures to reduce risk. Examples are prescribing burning to reduce fuel loads, breaking large plantations into smaller blocks to prevent fire spread, mixing species of lower and higher flammability to reduce overall fire risk, and ensuring that all parts of a plantation have access to fire suppression equipment. Planting a variety of tree species together so that no single host species is dominant can reduce insect outbreaks. Other disturbances (wind storms, earthquakes) are more difficult to guard against, but a healthy vigorous forest stand generally reduces the effects as much as possible.

- **Elimination of causes of reversal from land-use change to reduce risk.** Elimination requires identifying the drivers of land-use change and designing the project to address them. For example, if a region's coffee prices are rising, then a project in that region may incorporate a wood-products cooperative or another means of improving the residents' livelihood. Such a project may give landowners an income similar to what they might earn if they converted their project lands to coffee.

- **Contracts.** Contracts may provide incentives to project partners and/or landowners to use best-management practices or other approaches aimed at reducing the risk of reversal. For example, contracts could require landowners to pay for new seedlings and replant them if trees die, or they could be paid a bonus if the trees survive or the biomass exceeds an agreed amount after five or ten years.

Project developers should document any actions that they take to reduce the risk of reversibility and how they incorporated them into the estimates of a project's GHG reduction. They also should also indicate whether these actions will fully or only partially avoid the reversal of carbon storage.

11.3 For Any Residual Risk of Reversibility, Establish Mechanisms to Compensate for Reversals of Carbon Storage

After a project developer has identified, assessed, and taken action to reduce the reversibility of carbon storage, some risks may still remain, which may affect a project either during its life (or crediting period) or after its life (or the end of the crediting period). These residual risks should be documented, as well as any mechanisms put in place to compensate for this reversal of carbon storage. These mechanisms typically address any liabilities associated with the reversal of the project's GHG reduction. Of course, both GHG programs and any private contracts associated with the project must determine the exact nature of the liabilities. Risk management mechanisms generally include measures to obtain additional GHG reductions elsewhere.

Some mechanisms for compensating for any future loss of carbon storage benefits are:

- **Financial instruments,** such as insurance, forward contracts for carbon delivery at a future date, and options for the delivery of stored carbon tonnes. Traditional forest or crop insurance may be adapted to include GHG reductions. The instrument should explain how it works and state the time period over which it would operate.

- **Legal contracts,** such as purchase and sale contracts for forestry and agricultural GHG projects that contain clauses/processes for restitution when carbon stocks are released. Restitution may mean going to the market to buy compensatory (temporary or permanent) equivalents to replace the GHG reductions. The legal contract should describe how liability is apportioned among the different parties.

• **Portfolios or buffers** are a form of self-insurance by a single GHG project or coinsurance by a set of GHG projects. GHG projects can hold a proportion of the GHG reductions in reserve against the risk of reversal. A slightly modified approach bundles GHG projects, and only a portion of the GHG reductions are sold or reported, thereby hedging against the risk of natural events affecting one project but not the others. Portfolios or buffers should document the terms of the agreement, the GHG projects they contain, and the time period that the agreement covers. The size of the buffer or backup carbon reserve should be estimated conservatively.

11.4 Develop a Plan to Monitor Carbon Reversibility

The project's monitoring plan should contain indicators to determine whether carbon storage is continuing (see chapter 10, Monitoring and Quantifying GHG Reductions). Often these indicators are simpler than those needed to monitor the GHG reductions from the primary effects, as they need to determine only the continued existence of the particular land use or management practice and its acceptable maintenance. The frequency of monitoring varies according to project types and areas. The monitors may use low-resolution satellite imagery to identify potential losses of carbon from illegal logging, fires, or pest outbreaks, or they may use random field inspections. These indicators may then trigger more detailed monitoring.

Project developers should list the indicators, how they will be used to assess reversibility, and how frequently they will be monitored, as some risks may need to be monitored more often. Project developers would do well to check with the relevant GHG program, as it may explain how long project activities need to be monitored to ensure that the carbon is still being stored. The monitoring reports should state whether a loss has occurred and how it was or will be replaced and verify that action has been taken (e.g., a certification of the delivery or replacement tonnes by an insurer).

As a cross-check, a project's carbon reversibility management plan should contain

1. Reversible elements of the project's GHG reduction that
   • List the project's elements (including carbon pools) that pertain to storing carbon.

   • For each element contributing to carbon storage, assess the potential for its reversibility, including the uncertainty associated with these assessments.

   • If the carbon storage is not reversible, briefly explain why this is the case.

2. Actions to reduce or eliminate the reversibility of GHG reductions that
   • List any actions taken to reduce the risk of reversibility and how they were incorporated into the estimates of a project's GHG reduction.

   • State whether these actions will fully or only partially avoid the reversal of carbon storage.

3. The list of residual risks should include any risks that remain after the mechanisms to compensate for reversals of carbon storage have been put in place.

4. The monitoring plan should list the indicators of reversal, how they will be used to assess reversibility. It should list how frequently they will be monitored, as well as an explanation of this frequency. Finally, a list of the items to include in the monitoring reports, for example, whether a loss has occurred, how it was replaced, and whether the replacement has been verified should be included.

12 Reporting GHG Reductions and Total Carbon Stocks

Reporting requirements usually are determined by GHG programs and vary from program to program. But if no GHG program requirements are available, the following guidance, along with the requirements listed in the Project Protocol, will generate a transparent and complete report for LULUCF project activities.

The report should describe how the information was collected and how decisions were made regarding the

- Description of the project.
- GHG assessment boundary.
- Baseline carbon stocks and GHG removals for each project activity and primary effect.
- Estimated total carbon stocks, GHG removals and GHG reductions for the GHG project.
- Monitoring plan.
- Carbon reversibility management plan.
- Annual monitoring and GHG reduction reports.

12.1   Description of the Project

In addition to the general project description outlined in the Project Protocol, developers should name and briefly describe all the mandatory and voluntary programs in which they have registered GHG reductions or other activities on the land, to avoid double-counting GHG reductions and improve transparency (including all non-GHG land-use or management incentive programs, such as the U.S. Department of Agriculture's Conservation Reserve Program).

The report should contain maps of both the geographic area and the project site. It also should state the start date of the project and the date when GHG reductions are first generated, as well as the estimated date that all expected GHG reductions will be achieved. Since the life of LULUCF project activities may extend beyond the crediting period, project developers should consider the future of the carbon stocks and GHG removals and plan for their continued monitoring and calculation.

12.2   The GHG Assessment Boundary

As with other GHG projects, LULUCF project developers should report all primary and significant secondary effects resulting from each project activity. If the project has two project activities, they and their effects should be reported separately in order to maintain transparency and accuracy.

12.3   Baseline Carbon Stocks and GHG Removals for Each Project Activity and Primary Effect

A complete list of all the characteristics used to define the geographic area, temporal range, and final list of baseline candidates should be provided. When baseline candidates and the project activity are compared by their product or service—for example, certain forest management project activities supplying a specific product or service—a description of the product or service as well as the quantity produced should be provided.

For both procedures (project-specific and performance standard), project developers should report both the total and the per unit land area baseline carbon stocks and GHG removals, as well the degree of uncertainty of their estimates and how this uncertainty is accounted for (e.g., by discounting total reductions, overestimating baseline carbon stocks).

12.4   Estimated Total Carbon Stocks, GHG Removals, and GHG Reductions for the GHG Project

Project developers should list all the calculations they used to estimate the GHG reductions, including how the baseline carbon stocks and GHG removals were determined—for example, by direct measurement, default factors, or modeling—as well as any and all assumptions made using those quantification methods. In addition, project developers should report the total carbon stocks from the project activity.

12.5   Monitoring Plan

The report should include all the elements of the monitoring plan in the Project Protocol's chapter 10, Monitoring and Quantifying GHG Reductions, and the process for data storage and backup.

12.6   Carbon Reversibility Management Plan

Because LULUCF project activities are vulnerable to unexpected carbon stock losses (e.g., pest infestation, fire, illegal logging), project developers should create a plan to assess and mitigate carbon reversibility on the project site. The report should contain all the information, models, and sources used to determine the level of risk as well as the intended steps to mitigate it (for a complete list, see chapter 11, Carbon Reversibility Management Plan).

12.7   Annual Monitoring and GHG Reduction Reports

GHG projects submitted to GHG programs that grant credits to project developers before the actual GHG removals occur should be monitored and quantified annually to ensure that the GHG reductions are actually achieved. When possible, these reductions should be calculated or verified by a certified third party.

Nipawin Afforestation Project

This case study illustrates the application of part III of the Project Protocol and the LULUCF Guidance to a hypothetical GHG project, using both the project-specific and the performance standard procedures to estimate baseline removals. The sections in this example have the same numbers as those of the chapters in part II of the LULUCF Guidance. This case study is meant only as an illustration, however; the various sections of an actual project report may need additional details or justifications. Furthermore, owing to space limitations in this document, some of the supporting materials (e.g., excel spread sheets with calculations) and visuals have been excluded but can be found online. To see a full copy of this report, please visit the GHG Protocol web site at www.ghgprotocol.org.

The GHG project presented here is an afforestation project planting hybrid poplar trees near the town of Nipawin in Saskatchewan, Canada, with the trees providing feedstock to the Nipawin Ethanol Plant (NEP).

Chapter 3: Defining the GHG Assessment Boundary

3.1 IDENTIFYING THE PROJECT ACTIVITIES

The Project Site

The project site is located near the town of Nipawin, a small prairie community of approximately 4,300 people located in the Dark Gray soil zone of the Aspen Parkland Ecoregion in

Saskatchewan, Canada. Although this area is highly suitable for tree growth, traditionally either agricultural crops, such as grains and oil seeds, have been planted on the land surrounding Nipawin, or it has been used as pastureland. The project site covers 30,000 ha of agricultural land surrounding the physical location of the NEP.

The Project Activity

The project activity is changing the land use from nonforest agricultural land to forest, by planting hybrid poplar trees on 1500 ha, approximately one section, of land each year for twenty years. In accordance with this planting pattern, the first section of land will be planted in year 1 and harvested and replanted in year 21. This planting pattern will provide the NEP with enough feedstock from the project site to produce ethanol exclusively from woody cellulosic biomass. In turn, the NEP will produce 75 million L/yr of fuel-grade ethanol by gasifying 150,000 tonnes of biomass per year.

3.2 IDENTIFYING PRIMARY EFFECTS

The *primary effect* is the increased removals and storage of CO_2 by means of biological processes, particularly in trees and soil.

To estimate the magnitude of the primary effect, the changes in the carbon stored in all relevant carbon pools are considered, including living biomass, dead biomass, and soils. Above- and belowground living biomass and soil carbon will increase as a result of the project activity and, therefore, will be measured/estimated and monitored. Because dead biomass does not accumulate in significant amounts during the project activity, it is not considered. The tree biomass and soil carbon pools will be directly assessed and monitored.

As a result of the "even-flow wood supply" forest management being used, during which the amounts of land afforested and harvested after twenty years will become constant, the change in carbon stocks from one year to the next will become zero and theoretically will remain constant for the rest of the project's life. No GHG reductions will be sought once that new steady state on the land has been reached.

3.3 CONSIDERING ALL THE SECONDARY EFFECTS
Figure E.1 shows the activities that may result in secondary effects in each section of land planted.

3.3.1 ONE-TIME EFFECTS
The Nipawin Afforestation Project has three one-time effects:
- **Site preparation.** The preparation is cultivating each 1500-ha area in order to kill weeds and prepare the soil for planting trees each year for the twenty-one years of the project. A potential source of GHGs is the CO_2 from the fossil fuels combusted by mechanical equipment.

- **Harvesting the trees.** A 1500-ha site is harvested in year 21 after twenty years of growth. Because the project lasts for just twenty-one years, there is only one harvest during this time that emits CO_2 from the fossil fuels combusted by the harvesting equipment.

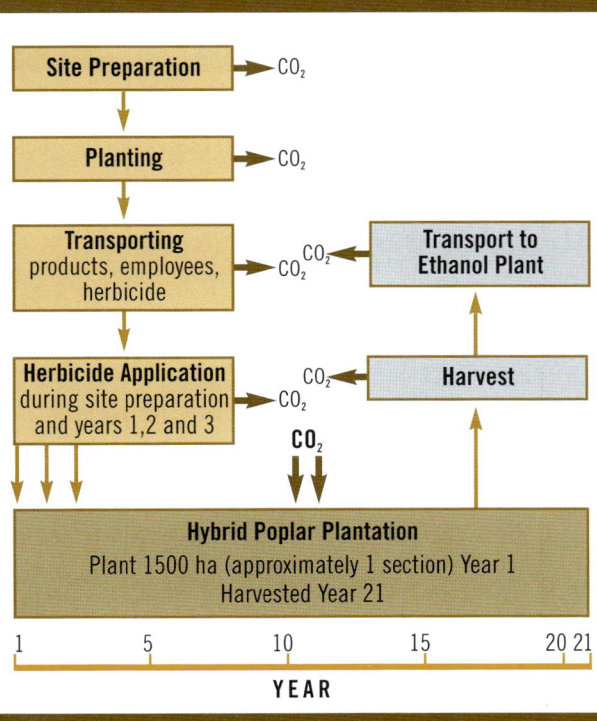

FIGURE E.1 Project-Related Activities per section that Result in GHG Emissions

- **Transportation.** The harvested trees are transported by truck to the NEP approximately two kilometers from the project site each year after year 21, resulting in CO_2 emissions from combusting fossil fuels.

3.3.2 UPSTREAM AND DOWNSTREAM EFFECTS
Possible upstream and downstream secondary effects are those from:
- **Nursery operations.** Nurseries provide planting stock for the afforested areas, associated with which are CO_2 emissions from heating and electricity for the greenhouses and from combusted fossil fuel emissions from transporting the planting stock.

- **Planting.** The hybrid poplar trees are planted by tractor, resulting in upstream effects from the CO_2 emissions related to diesel combustion.

- **Herbicide application.** Herbicides are applied to the hybrid poplar plantation while preparing the site and during the first three years after planting to control weeds and encourage tree growth. Therefore, herbicides are applied on 1500 ha in year 1, 3000 ha in year 2, 4500 ha in year 3, and 4500 ha in years 4 to 21. Potential GHG sources are the CO_2 emissions from fossil fuel use when the herbicides are manufactured and applied.

- **Transportation.** Products, employees, and herbicides are transported to and from the plantation, resulting in CO_2 emissions from the combustion of fossil fuels.

- **Market response 1.** There is a potential market response to the increased supply of wood in the region.

- **Market response 2.** Another market response may be provoked by removing land from agricultural use.

3.3.3 MITIGATING SECONDARY EFFECTS
Secondary effects were mitigated by
- Using fuel-efficient machinery, when and where possible, during the site preparation, for example, using diesel instead of gasoline vehicles.

- Not applying fertilizer.

Mitigating Market Responses
The amounts of both the land taken out of agricultural production and the increase in wood products for the fiber market were small, so it was determined that no actions were needed to mitigate the market responses.

3.4 ESTIMATING THE RELATIVE MAGNITUDE OF SECONDARY EFFECTS
Because most of the secondary effects were small compared with the primary effect, using default or existing data was determined to be the most cost-effective approach.

In addition, there was no market assessment, as the 30,000-ha project site will remove only 0.21 percent of

61

arable land from production, which is not likely to have a significant impact.

The percentage of arable land was determined by looking at the biophysical traits of the soils (Prairie Farm Rehabilitation Administration 2001) in the Dark Gray and Black soil zones[1] of Manitoba, Saskatchewan, and Alberta (*Soil Landscapes of Canada Working Group* 2006). The area of Dark Gray soil components in the prairies, 6,766,101 ha, and the area of Black soil components, 17.3 million ha are mostly represented by the Chernozems, Solonetzic, and Luvisols soils. Of the total area in the Dark Grey and Black soil zones (24,966,101 ha), 14,026,492 ha was determined to be arable (adapted from Statistics Canada 2001, see table E.1).

TABLE E.1 Commercial Cropland Area in the Dark Gray and Black Soil Zones of the Prairie Provinces, Canada

PROVINCE	COMMERCIAL CROPLAND LAND (ha)
Manitoba	3,946,254
Saskatchewan	6,794,027
Alberta	3,286,211
Total	**14,026,492**

Source: Adapted from Statistics Canada 2001.

3.5 ASSESSING THE SIGNIFICANCE OF SECONDARY EFFECTS

The secondary effects related to diesel fuel usage (i.e., site preparation, herbicide application, planting, and harvest), as well as herbicide manufacturing and transport, machinery manufacturing, and nursery operations, have been shown to be substantially smaller than the primary effect of afforestation (Heller, Keoleian, and Volk 2003, chap. 3, annex A). Box E.1 has a specific calculation for transporting the harvest in year 21 to double-check this assumption. Because the emissions from the transportation are minuscule compared with the GHG removals from the GHG project, the GHG emissions from transportation and other operations listed by Heller and colleagues are considered insignificant and are not included in the GHG assessment boundary. The emissions from the transportation of herbicides, employees, and equipment to the project site also are considered insignificant.

Market Responses

The potential market response from removing land from agricultural use also is considered insignificant. Although the project activity will remove 30,000 ha from agricultural production, the market response is expected to be small because the agricultural market is well developed and consumers can easily obtain alternative sources of the displaced grains and oil seeds.

The potential market response to increasing the wood supply is considered insignificant because the project is producing feedstock for energy production and will have no impact on any of the traditional wood markets (e.g., pulp for paper and wood for wood products).

BOX E.1 Transportation of Biomass to the NEP from the Plantation

The amount of carbon emitted from transporting biomass from 1500 ha in year 21 is found as follows:

Heavy-duty vehicles emit 1,094.78 g CO_2/mile, or 680.26 g CO_2/km, using gasoline diesel (0.030 percent sulfur) (S&T[2] Consultants Inc. 2005),

where

1 CO_2 unit = 0.2727 units of C (US EPA 2006).

Therefore, 185.51 g of C, or 0.000186 t C, are emitted for each kilometer traveled.

Because 1500 hectares contains 163,547 t C and the biomass content is twice the carbon content, the total biomass equals 163,547 x 2 = 327,094 t biomass.

Trucks can carry approximately 20 t of biomass per trip. Therefore, 327,094 t biomass will require 16,355 truck trips of approximately 2 km to the ethanol plant and 2 km back to the plantations (i.e., a 4-km round-trip).

To calculate the emissions from the transportation, use the following equations:

- 0.000186 t C · 4 km = 0.00074 t C

- 0.00074 t C · 16,355 truck trips = 12.14 t C released to transport the year 21 harvest to the NEP

The total GHG reduction for all 21 years of the Nipawin Afforestation Project is 1,879,612 t C (see section 10).

The emissions during transportation are therefore deemed insignificant.

EXAMPLE: Nipawin Afforestation Project

Chapter 4: Selecting a Baseline Procedure

This example compares the two methods of estimating baseline GHG removals, the project-specific procedure and the performance standard procedure.

Chapter 5: Identifying the Baseline Candidates

5.1 DEFINING THE PRODUCT OR SERVICE PROVIDED BY THE PROJECT ACTIVITY

The project activity is a change in land use, with different products produced by the project activity and the baseline candidates (see table E.2). The types of baseline candidates represent alternative land uses for the project site.

TABLE E.2 Alternative Land Uses and Their Products

ALTERNATIVE LAND USES	PRODUCT
Poplar plantation (project activity)	Wood fiber
Pastureland	Animal forage
Oil-seed and cereal cropland	Oils and grains

5.2 IDENTIFYING POSSIBLE TYPES OF BASELINE CANDIDATES

The possible types of baseline candidates are, broadly, the alternative practices or activities that could be undertaken on this project site and include

- **Commercial cropland with summer fallow:** Various types of crops are found on this land (table E.3). Summer fallow means letting land stand "idle" without a crop for a growing season in order to store soil moisture, control weed growth, and reduce the risk of crop failure in regions lacking moisture.

- **Commercial cropland without summer fallow.**

- **Pastureland:** Pastureland growing animal forage. The two types of pastureland in the region are tame and natural pasture. Disturbed land—for example, cropland, developed, or forested areas—cannot be converted to natural pasture, so natural pasture cannot be a baseline candidate for the project site.

- **Hybrid poplar forest:** The afforestation project in the region by Al-Pac.

- **Christmas tree production:** Spruce plantations that are harvested and sold for Christmas trees.

- **Commercial development:** The conversion of land to industrial or retail building structures for commercial production or retail.

TABLE E.3 Crops Constituting More Than 5% of Total Cropped Land (2001)

CROP	PERCENT OF CROPPED LAND
Wheat	30
Barley	17
Canola	16
Alfalfa	13
Oats	8

Source: Adapted from Statistics Canada 2001.

- **Residential development:** The conversion of land to residential building development.

- **Farmhouse development:** The conversion of land to farm infrastructure (e.g., farmhouses, sheds, barns).

5.3 DEFINING THE GEOGRAPHIC AREA AND THE TEMPORAL RANGE

5.3.1 DEFINING THE GEOGRAPHIC AREA

The following characteristics were considered when defining the geographic area:

- **Biophysical conditions.**
  - Suitability for hybrid poplar growth (Joss et al., 2005.), which corresponds to the Dark Gray and Black soil zones.
  - Ecoregion: Aspen Parkland (an ecoregion classified by the Saskatchewan Environment and Resource Management).
  - Slope: Classes 1 to 3, land with 1 to 5 percent slopes (i.e., near level to gentle slopes).
  - Climate: For Nipawin, a dry temperate climate.

- **Human-influenced factors.**
  No specific legal, zoning, or regulatory requirements affect the project site, although Saskatchewan contains publicly owned "Provincial Forest" that does not allow afforestation with exotic species such as poplar. In addition, the project site is located in a "Census Agricultural Region" and is defined as "Farm," information that is used to help refine the baseline candidates and the geographic area.

- **Availability of physical infrastructure.**
  - Roads: The entire region has many existing roads that are used for farm access.
  - Markets: The area has well-established markets for grains (i.e., the Wheat Pool), oil seeds, and wood as wood products, and there will be a market for ethanol biomass feedstocks once the NEP begins operating.

63

The geographic area for the Nipawin Afforestation Project was defined using the following characteristics (in this order):

- **Political boundaries:** The geographic area was restricted to Canada's prairie provinces (Alberta, Saskatchewan, and Manitoba), as they have the same general land-use conditions and agricultural and forest production laws and incentives.

- **Data boundaries:** The geographic area is limited to lands within Statistics Canada's Census Agricultural Regions, as they best represent the project site classification and facilitate the definition of the hectares in each baseline candidate for the performance standard procedure.

- **Biophysical conditions:** The geographic area was further restricted to the Dark Gray and Black soil zones, as they are highly suitable for growing hybrid poplars.[2]

- **Land tenure:** The geographic area was restricted as well to privately held lands, as the project activity will take place on private land. This means Provincial Forests are excluded.

In summary, the initial geographic area[3] is the privately held land on the Dark Gray and Black soil zones of Alberta, Saskatchewan, and Manitoba within the Census Agricultural Regions for these three provinces. This broad area is 21,916,393 ha, with the total amount of land in agricultural production (i.e., cropland and pastureland) in 2001 based on economical traits (Statistics Canada 2001)[4]. Table E.4 shows each agriculture land use as percentage of total agricultural land in the dark gray and black soil zones (2001).

5.3.2 DEFINING THE TEMPORAL RANGE

The area of agricultural land in commercial cropland remained relatively constant between 1996 and 2001 (table E.5, supporting material A online), although the conventional tillage management practices have recently shifted to zero tillage. There has also been a reduction in commercial cropland with summer fallow. There has been an increase in complementary and rotational grazing systems and perennial crop production (Boehm et al. 2004). These changes affect both the final list of baseline candidate and how the carbon stocks are estimated for these baseline candidates.

A five-year temporal range (1996 to 2001) will be used because the data for a longer period was not readily available. This is not ideal, as the temporal range should be at least ten years. But because this is only an illustrative example, it was not practical to purchase the additional historical data to extend the period. Also, because the land uses in the area are relatively stable, five years was not considered to be misleading.

TABLE E.4 Agriculture Land Use as Percentage of Total Agricultural Land in the Dark Gray and Black Soil Zones (2001)

LAND USE	TOTAL AGRICULTURAL LAND (%)
Commercial cropland (excluding Christmas trees/summer fallow area)	59
Natural pastureland (not a baseline candidate)	21
Tame or seeded pastureland	8
All other land (including Christmas tree area)	7
Commercial cropland with summer fallow land	5
Total	100

Source: Adapted from Statistics Canada 2001.

TABLE E.5 Change in Agricultural Land-Use Area from 1996 to 2001

LAND USE	CHANGE IN AREA (%)
Total agriculture	-1
Commercial cropland (excluding Christmas trees/summer fallow area)	1
Commercial cropland with summer fallow land	-36
Tame or seeded pastureland	12
All other land (including Christmas tree area)	-20

Source: Adapted from Statistics Canada 1996, 2001.

EXAMPLE: Nipawin Afforestation Project

5.4 DEFINING OTHER CRITERIA USED TO IDENTIFY THE TYPES OF BASELINE CANDIDATES

Alberta, Saskatchewan, and Manitoba have no stated legal requirements that affect land use in the geographic area, and no other criteria were identified to define the types of baseline candidates.

5.5 IDENTIFYING THE FINAL LIST OF BASELINE CANDIDATES

Four types of baseline candidates fall into the defined geographic area and temporal range:

1. **Baseline Candidate Type 1:** Commercial cropland. In 2001, 30 percent of the cropland in the geographic area was wheat, 17 percent was barley, 16 percent was canola, 13 percent was alfalfa, and 8 percent was oats, with the remainder consisting of miscellaneous crops types (Statistics Canada 2001). All crops are suitable for the commercial cropland baseline candidate. To be conservative in the quantification of the baseline carbon stocks, zero-tillage soil management is assumed.

2. **Baseline Candidate Type 2:** Tame pastureland. To be conservative in the quantification of the baseline carbon stocks, improved pasture management is assumed.

3. **Baseline Candidate Type 3:** Hybrid poplar forest.

4. **Baseline Candidate Type 4:** Development. Commercial, residential, and farmhouse development were aggregated.

Note that for the project-specific procedure, the types of baseline candidates represent the baseline candidates but that for the performance standard procedure, the individual baseline candidates are the individual hectares of land represented by each type of baseline candidate. The hectares are determined in section 7.1.2.

Based on several remote sensing and GIS maps from the late 1990s,[5] certain land-use changes were excluded from the list of baseline candidates. Because no land was converted from cropping or pasture to Christmas tree production during this time, Christmas tree production also was excluded.

Commercial cropland with and without summer fallow was aggregated to form a "commercial cropland" baseline candidate. The acreage under summer fallow decreased by 36 percent between 1996 and 2001 (table E.4) (Statistics Canada 1996, 2001), because summer fallow reduces the soil's organic matter relative to that of continuously cropped soils (Campbell et al. 2002), it is conservative to aggregate these two commercial cropping areas.

5.6 IDENTIFYING TYPES OF BASELINE CANDIDATES THAT REPRESENT COMMON PRACTICE

The common practice in the geographic area is commercial cropland, representing 64 percent of the geographic area.

Chapter 6: Estimating Baseline GHG Removals—Project-Specific Procedure

To estimate the baseline GHG removals using the project-specific procedure, the baseline scenario is identified; the change in carbon stocks for the baseline scenario are quantified; and the carbon stock changes are then converted to GHG removals (i.e., CO_2eq). In addition, the total carbon stocks are estimated for the baseline scenario (i.e., t C), in order to compare them with the project activity's total carbon stocks.

6.1 PERFORMING A COMPARATIVE ASSESSMENT OF BARRIERS

The possible alternatives for the baseline scenario—including the baseline candidates, the project activity, and the continuation of current activities—are evaluated by comparatively assessing the barriers. The possible alternatives for the baseline scenario (i.e., those identified in section 5.5) are:

- **Baseline Candidate 1:** Commercial cropland assuming zero-tillage. This represents the continuation of current activities.

- **Baseline Candidate 2:** Tame pastureland, assuming improved management of tame pasture.

- **Baseline Candidate 3:** The Al-Pac hybrid poplar forest.

- **Baseline Candidate 4:** Commercial, residential, and farmhouse development.

- **Project Activity:** The NEP hybrid popular forest.

Note that baseline candidate 3 and the project activity will be considered together in the comparative assessment of barriers, as the barriers are perceived to be similar in both cases, despite the differences between the projects.

6.1.1 IDENTIFYING BARRIERS TO THE PROJECT ACTIVITY AND BASELINE CANDIDATES

The anticipated barriers to the project activity and baseline candidates are listed next. Credible and justifiable data for each barrier category should be provided to support each claim. In this example, for the sake of brevity, the required supporting data (e.g., local labor statistics indicating the lack of possible maintenance personnel) are not provided.

The following barrier categories were considered:

1. **Financial and budgetary**
   - Commercial cropland: No financial or budgetary barriers were identified.

   - Tame pastureland: Financial and budgetary barriers were identified. Because most pastureland is marginal agricultural land, any conversion of commercial cropland to pasture would lower farm income.

   - Hybrid poplar forest: High perceived risks are barriers to this baseline candidate, first, the risk associated with reduced productivity due to fire and insect disease outbreaks[6] and, second, the risk related to the unproven business model and the long lag time between up-front expenditures and the revenue stream for tree plantations.

   - Development: A high perceived risk is a barrier to this baseline candidate. That is, commercial or residential development requires developers to purchase land from a private landowner (cost) as well as to assume the risk of these financial infra-structural investments. Farmhouse development requires a financial investment by the landowner, although the level of investment is not so high as that for commercial or residential development.

2. **Technology operation and maintenance**
   - Commercial cropland: No technology operation and maintenance barriers were identified.

   - Tame pastureland: No technology operation and maintenance barriers were identified.

   - Hybrid poplar forest: The lack of trained personnel is a barrier. Because hybrid poplar plantations are new in the region, there are few trained personnel capable of maintaining, operating, or managing such plantations. In addition, landowners do not know how to grow this tree crop species.

   - Development: The lack of maintenance personnel is a barrier. Land development requires equipment, building, infrastructure, and/or land maintenance, and the region has a limited maintenance workforce.

3. **Infrastructure**
   - Commercial cropland: No infrastructure barriers were identified.

   - Tame pastureland: No infrastructure barriers were identified.

   - Hybrid poplar forest: No infrastructure barriers were identified.

   - Development: Infrastructure barriers were identified for development because building supplies would

need to be sourced and purchased, and entire new structures would need to be built. In addition, the potential impact of development in the project area (i.e., 30,000 ha) would be minimal because the amount of development relative to the size of the project area would be negligible.

4. **Market structure**
   - Commercial cropland: No market structure barriers were identified.

   - Tame pastureland: No market structure barriers were identified. Pastureland is either used by the landowner or rented to neighboring farmers.

   - Hybrid poplar forest: Market structure barriers do exist, as there is currently no market for woody biomass in the region. However, the Nipawin Ethanol Plant is expected to buy all biomass produced by the poplar plantations for ethanol feedstock.

   - Development: Markets related to commercial, residential, and farmhouse developments are relatively unknown. In general, however, the area's population has not been expanding quickly, so there is not a large demand for new residential areas or commercial goods and, therefore, development.

5. **Institutional/social/cultural/political**
   - Commercial cropland: No institutional, social, cultural, or political barriers were identified.

   - Tame pastureland: No social, cultural, etc., barriers were identified.

   - Hybrid poplar forest: Social and cultural barriers were identified. Because residents of the area have been commercially growing crops for the last one hundred years, some of them are resistant to converting agricultural land to poplar plantations. Furthermore, in the past, agricultural soils classified as Dark Gray Luvisols were cleared of trees for agriculture, so landowners are reluctant to plant trees on land that their ancestors worked hard to clear. There is also a public perception that tree plantations compete with food production.

   - Development: Potential institutional and political barriers were identified. Any commercial or residential development must comply with zoning laws.

6. **Resource availability**
   - Commercial cropland: No resource availability barriers were identified.

   - Tame pastureland: No resource availability barriers were identified.

   - Hybrid poplar forest: Barriers to resource availability were identified, that is, concerns about seedling availability and nursery capacity in meeting the demand for planting stock.

EXAMPLE: Nipawin Afforestation Project

- Development: No resource availability barriers were identified.

6.1.2 IDENTIFYING BARRIERS TO THE CONTINUATION OF CURRENT ACTIVITIES

Commercial cropland represents the continuation of current activities, and no barriers were identified that would affect commercial cropland. Furthermore, no legal or market shifts were expected that would affect commercial cropland.

6.1.3 ASSESSING THE RELATIVE IMPORTANCE OF THE IDENTIFIED BARRIERS

The relative importance of barriers was assessed for each alternative. Overall, the financial/budgetary and social/cultural barriers were the most significant barriers identified, with the technology operation and maintenance and market structure barriers considered less important than the other barriers.

Table E.6 ranks the importance of the barriers for each baseline scenario alternative, indicating that commercial cropland and tame pastureland face the lowest barriers.

6.2 IDENTIFYING THE BASELINE SCENARIO
6.2.1 EXPLAINING BARRIERS TO THE PROJECT ACTIVITY AND HOW THEY WILL BE OVERCOME

The project activity, hybrid poplar forest, faces five types of barriers: financial/budgetary, technology operation and maintenance, market structure, social/cultural, and resource availability. For the GHG project to succeed, the following measures were implemented:

1. Financial and budgetary barriers
- Innovative financing arrangements were offered to offset the risks associated with the long lag time between up-front expenditures and the revenue stream for forest projects. For example, farmland will be leased from the landowners, paying them an annual fee based on market rates. In some situations, the landowners will also be hired to tend the plantation during its growth cycle.

- The plantations were located in areas with fire protection to minimize the risk of fire.

- New uses were created for biomass (e.g., the NEP).

- "Due diligence and risk assessments" were incorporated in the business model to reassure investors of the project's success.

2. Technology operation and maintenance barriers
- Silviculture management training (e.g., on-site silviculture training and classes covering silviculture theory, growth curves, and calculations) was offered to plantation workers through community colleges and with forestry experts.

- Partnerships were created with forestry experts to support education and nursery production.

- Home study courses were provided for landowners on tree crop production.

3. Market structure barriers
- The viability of the popular plantations is integrally linked to the success of the Nipawin Ethanol Plant, as it is the sole market for the plantations' wood products. The Nipawin Afforestation Project is not, however, engaged in any activities aimed at ensuring the plant's viability.

4. Institutional, social, cultural, and political barriers
- To improve the public perception of the project, promotional material demonstrating that poplar plantations are a viable land-use alternative that is environmentally sustainable and economically beneficial and improves farm income security will be developed and distributed to the local community. This material will show the environmental benefits of establishing the plantations, such as improvement of the soil, water, and habitat. In addition, field tours of the project site will be conducted annually during the first five years of the project.

5. Resource availability barriers
- Stool beds—densely planted trees grown specifically to supply cuttings for plantations—will be established on the participating landowners' land.

6.2.2 IDENTIFYING THE BASELINE SCENARIO USING THE COMPARATIVE ASSESSMENT OF BARRIERS

Tables E.6 and E.7 show that the project activity has the highest barriers; development has medium barriers; tame pastureland has low barriers; and commercial cropland has no barriers. Therefore, commercial cropland is identified as the baseline scenario.

6.2.3 JUSTIFYING THE BASELINE SCENARIO

The comparative assessment of barriers identified commercial cropland as the baseline scenario. Commercial cropland represents the continuation of current activities on the project site and is the common practice in the geographic area.

Because the comparative assessment of barriers clearly identified a baseline scenario, it was not necessary to conduct a net benefits assessment.

6.3 ESTIMATING THE BASELINE GHG REMOVALS AND TOTAL CARBON STOCKS

Baseline Carbon Stock Changes and GHG Removals

The baseline carbon stocks changes were estimated for commercial cropland using the following information and data. Default values were used for carbon stock data. The carbon pools included in the carbon stock estimations were the living biomass (i.e., above- and belowground biomass) and soils. These pools are expected to increase from afforesting cropland. Because dead biomass is not expected to accumulate to any large degree in poplar forests or on commercial cropland, it was excluded.

Living Biomass
Carbon stocks register no aboveground or belowground change over time on commercial cropland, as vegetative matter is harvested and removed each year (IPCC 2003).

Therefore, the change in living biomass carbon
$$= 0 \text{ t C/ha/yr.}$$

Soils
The accumulation of carbon from zero-tillage practices on commercial cropland in the Black and Dark Gray soil zones is 0.37 t C/ha/yr (Boehm et al. 2004), with soil organic carbon levels reaching a steady state after approximately twenty years.

Therefore, assuming the soils are in equilibrium, the change in soil carbon over time is 0 t C/ha/yr.

The change in carbon stocks on commercial cropland
$$= \text{change in living biomass} + \text{change in soil carbon}$$
$$= 0 \text{ t C/ha/yr} + 0 \text{ t C/ha/yr}$$
$$= 0 \text{ t C/ha/yr}$$

The baseline GHG removals/ha/yr
$$= 0 \text{ t C/ha/yr} \cdot \tfrac{44}{12} \text{t CO}_2/ \text{ t carbon}$$
$$= 0 \text{ t CO}_2\text{/ha/yr}$$

Total Baseline Carbon Stocks
To estimate the total carbon stocks stored for commercial cropland, the change in carbon stock each year is added to the equilibrium carbon stocks for the project lifetime. Because total baseline carbon stock information is not used in calculating the GHG reduction, this information need not be translated to t CO$_2$ unless specifically required by a GHG program.

Equilibrium carbon stocks refer to the carbon stored at time zero of the project's implementation and is assumed to remain constant over time as long as there is no change in land use or management. Because living biomass does not accumulate on commercial cropland, only soil carbon is included in the equilibrium carbon stocks.

The equilibrium soil carbon stocks were estimated using the IPCC's methods for calculating soil organic carbon (SOC) at the beginning of the inventory period (IPCC 2003).

$$SOC = SOC_{REF} \cdot F_{LU} \cdot F_{MG} \cdot F_I$$

where

SOC	= soil organic carbon in the inventory year, t C/ha
SOC_{REF}	= the reference soil carbon stock, t C/ha
F_{LU}	= carbon stock change factor for land use or land-use change type, dimensionless[7]
F_{MG}	= carbon stock change factor for management regime, dimensionless
F_I	= carbon stock change factor for input of organic matter, dimensionless

For a Chernozem in the Black soil zone or a Luvisol in the Dark Gray soil zone in a cold, dry temperate climate, SOC_{REF} is 50 t C/ha. Based on one hectare of land under zero-tillage long-term annual cropping with medium carbon input levels, the soil organic carbon at the beginning of the inventory period is

$$\text{SOC} = 50 \text{ t C/ha} \cdot 0.82 \cdot 1.10 \cdot 1.0[8]$$
$$= 45.10 \text{ t C/ha}$$

Therefore, the equilibrium carbon stock = 45.10 t C/ha.

Each year the change in carbon stock for that year is the new carbon stored in the biological system. For each time period the carbon stock per hectare is the equilibrium carbon stock plus all the carbon stored from previous years. The annual change in carbon stocks is added to the equilib-

TABLE E.6 Cumulative Importance of Barriers for Each Baseline Scenario

BASELINE SCENARIO	BARRIER 1: FINANCIAL/ BUDGETARY (H)	BARRIER 2: TECHNOLOGY O&M (M)	BARRIER 3: INFRA- STRUCTURE (M)	BARRIER 4: MARKET STRUCTURE (L)
Baseline Candidate 1: Commercial cropland (continuation of current activities)	Not present	Not present	Not present	Not present
Baseline Candidate 2: Tame pastureland	Low	Not present	Not present	Not present
Baseline Candidate 3: Hybrid poplar forest (project activity)	High	Medium	Not present	Medium
Baseline Candidate 4: Development	High	Medium	Low	Low

Note: The relative importance of the barriers compared with one another: H = significant barrier, M = moderately significant barrier, L = less significant barrier.

EXAMPLE: Nipawin Afforestation Project

TABLE E.7 Results of Comparative Assessment of Barriers

BASELINE SCENARIO ALTERNATIVES	RANK BY CUMULATIVE IMPACT OF BARRIERS	CONCLUSION
Baseline Candidate 1: Commercial cropland (continuation of current activities)	No barriers	Accept as baseline scenario
Baseline Candidate 2: Tame pastureland	Low barriers	Reject as baseline scenario
Baseline Candidate 3: Hybrid poplar forest (project activity)	Highest barriers	Reject as baseline scenario
Baseline Candidate 4: Development	Medium barriers	Reject as baseline scenario

rium carbon stock to get the total carbon stock for commercial cropland (see table E.8). The total baseline carbon stocks (t C) is the carbon stocks per hectare multiplied by the land base (i.e., 30,000 ha).

Chapter 7: Estimating the Change in Baseline GHG Removals— Performance Standard Procedure

7.1 TIME-BASED PERFORMANCE STANDARD

7.1.1 SPECIFY THE APPROPRIATE PERFORMANCE METRIC

The performance metric:

$$\frac{\text{GHG removals}_t}{\text{unit area of land}}$$

7.1.2 CALCULATE GHG REMOVALS FOR EACH BASELINE CANDIDATE IN EACH TIME PERIOD

Identifying Total Hectares

For the performance standard, baseline candidates are the individual hectares of each type of baseline candidate. Data from both Statistics Canada (2001) and *Soil Landscapes of Canada Working Group* (2006) are used to define the hectares attributable to each baseline candidate type.

BARRIER 5: CULTURAL/ SOCIAL/ETC.(H)	BARRIER 6: RESOURCE AVAILABILITY (M)	RANK BY CUMULATIVE IMPACT
Not present	Not present	No barriers
Not present	Not present	Low barriers
High	Medium	Highest barriers
Low	Not present	Medium barriers

- *Commercial cropland:* In 2001, crops were grown on 64 percent of agricultural land (see table E.4). The project's geographic area contains 14,026,492 ha of arable, commercial cropland (Statistics Canada 2001).

- *Tame pastureland:* The geographic area also contains approximately 1,753,311 ha of tame pastureland, based on the area of pastureland located in the relevant (i.e., Dark Gray and Black soil zones) Census Agricultural Regions (Statistics Canada 2001).

- *Hybrid poplar forest:* The geographic area contains only 3000 ha of hybrid poplar forest, the Al-Pac plantation.[9] This plantation is new and has been converting cropland to forest since 2001.

- *Development:* This baseline candidate represents such a small number of hectares in the geographic area that it cannot have a significant impact on the calculation of the performance standard and therefore is not quantified.

The total geographic area for the performance standard is 15,782,803 ha.

Baseline Carbon Stock Changes and GHG Removals

Default carbon stock data were used to estimate the annual baseline carbon stocks changes for the commercial cropland and tame pastureland. The growth curve for hybrid poplar, the total root biomass prediction equation (Li et al. 2003), and default carbon stock data were used to estimate the annual baseline carbon stock changes from the hybrid poplar forest baseline candidates.

When using default data, it is not possible to estimate the specific carbon stock changes for each baseline candidate. Therefore, representative carbon stock changes were derived for each type of baseline candidate, and this value was extrapolated to all individual baseline candidates.

The carbon pools included in the carbon stock changes estimate were living biomass (above- and belowground biomass) and soil carbon. Both pools increase as a result of afforestation. Because dead biomass does not appreciably increase on commercial cropland, pastureland, or hybrid poplar forestland, it was excluded.

TABLE E.8 Total Carbon Stocks for Commercial Cropland		
YEAR	**CARBON STOCKS PER HECTARE**	**TOTAL BASELINE CARBON STOCK**
	t C/ha	t C
	45.10	1,353,000
1	45.10	1,353,000
2	45.10	1,353,000
3	45.10	1,353,000
4	45.10	1,353,000
5	45.10	1,353,000
6	45.10	1,353,000
7	45.10	1,353,000
8	45.10	1,353,000
9	45.10	1,353,000
10	45.10	1,353,000
11	45.10	1,353,000
12	45.10	1,353,000
13	45.10	1,353,000
14	45.10	1,353,000
15	45.10	1,353,000
16	45.10	1,353,000
17	45.10	1,353,000
18	45.10	1,353,000
19	45.10	1,353,000
20	45.10	1,353,000
21	45.10	1,353,000

Baseline Candidate 1: Commercial Cropland
The commercial cropland GHG removals and carbon stocks were estimated as part of the project-specific procedure, with table E.8 showing the total carbon stocks.

Baseline Candidate 2: Tame Pastureland
Baseline Carbon Stock Changes and GHG Removals
Living Biomass
The aboveground component of the living biomass carbon pool is small, and both the above- and belowground components are relatively insensitive to management (IPCC 2003). Therefore, it is assumed that there will be no changes in living biomass carbon stocks over time (IPCC 2003).

Soils
The carbon stock for soil with improved grazing management in the Aspen Parkland Ecoregion is 0.09 t C/ha/yr for tame pastureland (Boehm et al. 2004). Soil organic carbon levels reach equilibrium after about twenty years, and the soils are assumed to be in equilibrium.

Therefore assuming the soils are in equilibrium, the change in soil carbon over time is 0 t C/ha/yr.

The change in carbon stocks on tame pastureland
= change in living biomass + change in soil carbon

= 0 t C/ha/yr + 0 t C/ha/yr

= 0 t C/ha/yr

The baseline GHG removals from tame pastureland
= change in carbon stocks $\cdot \frac{44}{12}$ t CO_2/t carbon

= 0 t CO_2/ha/yr

Total Baseline Carbon Stocks
To estimate the total carbon stocks for tame pastureland, the carbon stock changes are added to the equilibrium carbon stocks.

Because living biomass does not accumulate on pastureland (i.e., carbon accumulation through plant growth is offset by losses through fire and decomposition), only soil carbon is included in the equilibrium carbon stocks.[10]

The equilibrium soil carbon stocks were estimated using the IPCC's methods for calculating soil organic carbon (SOC) at the beginning of the inventory period (IPCC 2003).

$$SOC = SOC_{REF} \cdot F_{LU} \cdot F_{MG} \cdot F_I$$

where

SOC = soil organic carbon in the inventory year given, t C/ha

SOC_{REF} = the reference soil carbon stock, t C/ha

F_{LU} = carbon stock change factor for land use or land-use change type, dimensionless[11]

F_{MG} = carbon stock change factor for management regime, dimensionless

F_I = carbon stock change factor for input of organic matter, dimensionless

For a Chernozem (Black soil zone) or a Luvisol (Dark Gray soil zone) in a cold, dry temperate climate, SOC_{REF} is 50 t C/ha. Based on one hectare of land in improved grassland with nominal inputs, the soil organic carbon at the beginning of the inventory period is

SOC = 50 t C/ha $\cdot$ 1.0 $\cdot$ 1.1 $\cdot$ 1.0^{12}
= 55.0 t C/ha

Therefore, the equilibrium carbon stock = 55.0 t C/ha

EXAMPLE: Nipawin Afforestation Project

The annual carbon stock changes are added to the equilibrium carbon stock to get the carbon stocks per hectare for pastureland (see table E.9). The total baseline carbon stocks (t C) is the carbon stocks per hectare multiplied by the land base (i.e., 30,000 ha).

Baseline Candidate 3: Hybrid Poplar Forest
Baseline Carbon Stock Changes and GHG Removals
Living Biomass
Aboveground carbon stocks were estimated by converting the biomass from the hybrid poplar growth curve (Peterson et al. 1999) to carbon (i.e., by multiplying biomass by 0.5). From this curve the change in carbon stocks also was estimated (table E.10). The average equilibrium or steady-state carbon stocks of 109 t C/ha is reached after twenty years, because an equivalent area is planted and harvested each year (see supporting material B online).

The belowground biomass was estimated using the total root biomass prediction equation developed for the region's hardwood species (Li et al. 2003):

$$\text{Belowground biomass} = 1.576 \cdot (\text{aboveground biomass})^{0.615}$$

This equation uses aboveground carbon stocks instead of aboveground biomass to estimate belowground carbon stocks (table E.10). As with the aboveground carbon stocks, the belowground carbon stocks reach a steady state of 28 t C/ha after twenty years (see supporting material B online).

Soils
Soil carbon increases by 0.5 t C/ha/yr during the first twenty years following afforestation (Niu and Duiker 2006). Then after twenty years, the soil carbon stocks reach an equilibrium of 10 t C/ha, and there is no new carbon stored in this pool.

Total Baseline Carbon Stocks
Finally, the baseline carbon stored from the afforestation activity is calculated by summing the above- and belowground baseline carbon stock and the soil carbon stock change (table E.10). After twenty years, the carbon stock reaches an equilibrium of 120 t C/ha (see supporting material B online).

Equilibrium carbon stocks are the carbon stocks stored when the land use is in a steady state, before the implementation of the project activity or another land-use change. Before the Al-Pac afforestation, the land was used for crop production. Because this project is still in its first years of implementation however, carbon is not in equilibrium and the GHG removals reflect the change in land use from cropland to forest, rather than a forest in a steady state.

The amount of carbon already present in the soil is the same as the preceding project-specific commercial cropland carbon stock calculation.

$$\text{SOC} = 50 \text{ t C/ha} \cdot 0.82 \cdot 1.10 \cdot 1.0^{15}$$

$$= 45.10 \text{ t C/ha}$$

The afforestation carbon stock changes are added to the initial cropland carbon stock to get the carbon stock for afforestation (see table E.11). The carbon stock value reaches an equilibrium at 165 t C/ha after year 20. The total carbon stock is found by multiplying the carbon stock per hectare by 30,000 ha.

71

TABLE E.9 Total Carbon Stocks for Pastureland

YEAR	CARBON STOCKS PER HECTARE	TOTAL BASELINE CARBON STOCK
	t C/ha	t C
0	55.0	1,650,000
1	55.0	1,650,000
2	55.0	1,650,000
3	55.0	1,650,000
4	55.0	1,650,000
5	55.0	1,650,000
6	55.0	1,650,000
7	55.0	1,650,000
8	55.0	1,650,000
9	55.0	1,650,000
10	55.0	1,650,000
11	55.0	1,650,000
12	55.0	1,650,000
13	55.0	1,650,000
14	55.0	1,650,000
15	55.0	1,650,000
16	55.0	1,650,000
17	55.0	1,650,000
18	55.0	1,650,000
19	55.0	1,650,000
20	55.0	1,650,000
21	55.0	1,650,000

TABLE E.10 Baseline Carbon Stocks and GHG Removals for Hybrid Poplar Forest

YEAR	CUMULATIVE INCREASE IN ABOVEGROUND CARBON STOCKS	CUMULATIVE INCREASE IN BELOWGROUND CARBON STOCKS	CUMULATIVE INCREASE IN SOIL CARBON STOCKS	TOTAL CUMULATIVE INCREASE IN CARBON STOCKS FROM ALL POOLS	CHANGE IN CARBON STOCKS	GHG REMOVALS
	t C/ha	t C/ha	t C/ha	t C/ha	t C/ha	t CO_2/ha
1	0.31	0.04	0.5	0.85	0.85	3.12
2	2.00	0.11	1.0	3.11	2.26	8.29
3	5.52	0.21	1.5	7.23	4.12	15.11
4	10.76	0.31	2.0	13.07	5.84	21.41
5	17.34	0.42	2.5	20.26	7.19	26.36
6	24.87	0.52	3.0	28.39	8.13	29.81
7	32.93	0.63	3.5	37.06	8.67	31.79
8	41.18	0.71	4.0	45.89	8.83	32.38
9	49.37	0.79	4.5	54.66	8.77	32.16
10	57.28	0.87	5.0	63.15	8.49	31.13
11	64.79	0.94	5.5	71.23	8.08	29.63
12	71.81	1.00	6.0	78.81	7.58	27.79
13	78.29	1.06	6.5	85.85	7.04	25.81
14	84.21	1.10	7.0	92.31	6.46	23.69
15	89.57	1.15	7.5	98.22	5.91	21.67
16	94.40	1.18	8.0	103.58	5.36	19.65
17	98.73	1.22	8.5	108.45	4.87	17.86
18	102.58	1.25	9.0	112.83	4.38	16.06
19	106.01	1.27	9.5	116.78	3.95	14.48
20	109.03	1.29	10.0	120.32	3.54	12.98
21	109.03	1.29	10.0	120.32	0	0

7.1.3 CALCULATE THE GHG REMOVALS FOR DIFFERENT STRINGENCY LEVELS

The *time-based performance standard stringency level* refers to how high the baseline GHG removals are relative to the GHG removals of all the baseline candidates. The stringency level is chosen by comparing the

- Most stringent GHG removals.

- Weighted mean GHG removals.

- Median GHG removals.

- GHG removals relating to two different percentiles that are at least better than average.

Most Stringent GHG Removals

Because neither the commercial cropland nor the pasture-land has any change in its GHG removals over time, the most stringent GHG removals are those for hybrid poplar forest (see table E.12).

EXAMPLE: Nipawin Afforestation Project

Weighted Mean GHG Removals

$$\text{Weighted mean GHG removals}_{jt} = \frac{\sum_{j=1}^{n} (CO_2 \text{ removals}_{jt} \cdot \text{area}_j)}{\sum_{j=1}^{n} (\text{area}_j)}$$

where

GHG removals$_{jt}$ = GHG removals for baseline candidate j in time period t

area$_j$ = area encompassed by baseline candidate j (e.g., 1 hectare)

n = total number of baseline candidates

j = individual baseline candidate

The formula was applied to the three baseline candidates—commercial cropland, pastureland, and hybrid poplar forest—to calculate the weighted mean GHG removals for years 1 to 21 (table E.13).

TABLE E.11 Total Carbon Stocks for Hybrid Poplar Baseline Candidate

YEAR	CARBON STOCK PER HECTARE	TOTAL BASELINE CARBON STOCK
	t C/ha	t C
0	45.10	1,353,000
1	45.95	1,378,500
2	48.21	1,446,300
3	52.33	1,569,900
4	58.17	1,745,100
5	65.36	1,960,800
6	73.49	2,204,700
7	82.16	2,464,800
8	90.99	2,729,700
9	99.76	2,992,800
10	108.25	3,247,500
11	116.33	3,489,900
12	123.91	3,717,300
13	130.95	3,928,500
14	137.41	4,122,300
15	143.32	4,299,600
16	148.69	4,460,700·
17	153.55	4,606,500
18	157.93	4,737,900
19	161.88	4,856,400
20	165.42	4,962,600
21	165.42	4,962,600

Median GHG Removals and GHG Removals Relating to Different Percentiles

For each year from 1 to 21, the baseline GHG removals of each baseline candidate were sorted from lowest to highest.

The GHG removals corresponding to a specific percentile *(pc)* is determined by
• Calculating its approximate rank, w

$$w = (a \cdot pc)/100 + 0.5$$

where

a = the total land area represented in the geographic area

• Assigning g to be the integer part of w and f the fraction part of w. Table E.14 lists the values for the 50th, 75th, and 90th percentiles.

• Calculating the GHG removal *(pe)* of the specific percentile *(pc)* using the following equation:

$$pe = (1 \text{-} f) \cdot x_g + f \cdot x_{g+1}$$

where

x_g = the GHG removals assigned to land unit g (table E.15)

7.1.4   SELECT AN APPROPRIATE STRINGENCY LEVEL

The stringency level selected is the weighted mean; the 50th, 75th, and the 90th percentiles are below average and cannot be chosen. The most stringent GHG removals—those equal to the Al-Pac GHG removals—would apply only to the 99.977 percentile, which would be very stringent and not appropriate for this example.

7.1.5 ESTIMATE THE BASELINE GHG REMOVALS AND TOTAL BASELINE CARBON STOCKS

The baseline GHG removals are equal to the weighted mean GHG removals, found in table E.13.

To find total baseline carbon stocks: first, calculate the percent of the geographic area that each type of baseline candidate represents; second, apply the percentage to the area of the project site; third, multiply the baseline carbon stocks by their respective area of the project site (see table E.16).

The total baseline carbon stocks for the performance standard are found in table E.17.

Chapter 8: Applying a Land-Use or Management Trend Factor

The *land-use trend factor* estimates the underlying rate at which the land use or management is changing in the geographic area. This factor is then used to adjust the baseline GHG removals to reflect any inherent trends in the area.

8.1 WHEN TO APPLY THE LAND-USE OR MANAGEMENT TREND FACTOR

Because the project activity and baseline candidates represent discrete land uses, a trend factor might be applied. But it was determined that the land-use trend factor was not

TABLE E.12 Most Stringent Stringency Level, Years 1 to 21

YEAR	BASELINE CANDIDATE 1 COMMERCIAL CROPLAND GHG REMOVALS t CO_2/ha	BASELINE CANDIDATE 2 TAME PASTURELAND GHG REMOVALS t CO_2/ha	BASELINE CANDIDATE 3 AL-PAC HYBRID POPLAR FOREST GHG REMOVALS t CO_2/ha	MOST STRINGENT GHG REMOVALS GHG REMOVALS t CO_2/ha
1	0	0	3.12	3.12
2	0	0	8.29	8.29
3	0	0	15.11	15.11
4	0	0	21.41	21.41
5	0	0	26.36	26.36
6	0	0	29.81	29.81
7	0	0	31.79	31.79
8	0	0	32.38	32.38
9	0	0	32.16	32.16
10	0	0	31.13	31.13
11	0	0	29.63	29.63
12	0	0	27.79	27.79
13	0	0	25.81	25.81
14	0	0	23.69	23.69
15	0	0	21.67	21.67
16	0	0	19.65	19.65
17	0	0	17.86	17.86
18	0	0	16.06	16.06
19	0	0	14.48	14.48
20	0	0	12.98	12.98
21	0	0	0	0
Number of hectares	14,026,492	1,753,311	3,000	

EXAMPLE: Nipawin Afforestation Project

applicable to this project activity because the temporal range was insufficient to estimate a reliable land-use trend factor. To ascertain the validity of this decision and to decide whether additional data should be collected, the land-use trends between 1996 and 2001 were assessed. As there was no appreciable conversion of land to forest during that period, no additional data were collected. Although the Al-Pac plantations were planted in 2001, they were so small compared with the geographic area that they were not considered relevant. The geographic area should be monitored for an increase of afforestation activities.

Chapter 10: Monitoring and Quantifying the GHG Reductions

10.2 QUANTIFYING GHG REDUCTIONS

Project Activity's Change in Carbon Stocks, GHG Removals, and Total Carbon

Change in Carbon Stocks and GHG Removals
Carbon stocks for the Nipawin Afforestation Project are based on 1500 ha of land being afforested each year and were estimated as follows:

Living Biomass
Aboveground carbon stocks were estimated by converting the biomass from the hybrid poplar growth curve (Peterson et al. 1999) to carbon. The carbon stocks reach an equilibrium of 57 t C/ha after twenty years (see table E.18 and supporting material C online). Box E.2 gives a sample calculation.

The belowground biomass was estimated using the following equation to predict the total root biomass for hardwood tree species (Li et al. 2003):

Belowground biomass = 1.576 • (aboveground biomass)$^{0.615}$

Aboveground carbon was used in the equation instead of aboveground biomass to obtain an estimate of belowground carbon. After twenty years, equilibrium carbon stocks are reached: 0.36 t C/ha for the project site, as shown in table E.18 and supporting material C online.

Soils
The soil carbon is based on 1500 ha of land being afforested each year for twenty years. Afforested land gains 0.5 t C ha/yr (Niu and Duiker 2006) (supporting material D). Box E.3 gives a sample calculation.

TABLE E.13 Weighted Mean GHG Removals for Commercial Cropland, Pasture, and Hybrid Poplar, Year 1 to 21.

YEAR	WEIGHTED MEAN GHG REMOVALS t CO_2/ha
1	0.00059
2	0.00158
3	0.00287
4	0.00407
5	0.00501
6	0.00567
7	0.00604
8	0.00615
9	0.00611
10	0.00592
11	0.00563
12	0.00528
13	0.00491
14	0.00450
15	0.00412
16	0.00374
17	0.00339
18	0.00305
19	0.00275
20	0.00247
21	0

75

TABLE E.14 Values Assigned to *w*, *g*, and *f* Variables for the 50th, 75th, and 90th GHG Removal Percentiles

VARIABLE	50TH PERCENTILE	75TH PERCENTILE	90TH PERCENTILE
w	7,891,402	11,837,102.25	14,204,522.7
g	7,891,402	11,837,102	14,204,522
f	0	0.25	0.70

TABLE E.15 GHG Removals for the 50th, 75th, and 90th Percentiles, Years 1 to 21

YEAR	RANK, m, FOR EACH ha			50TH PERCENTILE	75TH PERCENTILE	90TH PERCENTILE
	1 TO 14,026,491	14,026,492 TO 15,779,802	15,779,803 TO 15,782,803			
	t CO_2/ha	t CO_2/ha	t CO_2/ha	t CO_2/ha	t CO_2/ha	t CO_2/ha
1	0	0	3.12	0	0	0
2	0	0	8.29	0	0	0
3	0	0	15.11	0	0	0
4	0	0	21.41	0	0	0
5	0	0	26.36	0	0	0
6	0	0	29.81	0	0	0
7	0	0	31.79	0	0	0
8	0	0	32.38	0	0	0
9	0	0	32.16	0	0	0
10	0	0	31.13	0	0	0
11	0	0	29.63	0	0	0
12	0	0	27.79	0	0	0
13	0	0	25.81	0	0	0
14	0	0	23.69	0	0	0
15	0	0	21.67	0	0	0
16	0	0	19.65	0	0	0
17	0	0	17.86	0	0	0
18	0	0	16.06	0	0	0
19	0	0	14.48	0	0	0
20	0	0	12.98	0	0	0
21	0	0	0	0	0	0

Project Activity's Total Carbon Stocks

Finally, the project's carbon stocks are calculated by summing the above- and belowground carbon stock and the soil carbon stock change (see table E.18). After twenty years, the carbon stock reaches a steady state of 62.66 t C/ha.

Commercial cropland is the current land use on the project site, so all additional carbon stored from the project activity has been added to the soil carbon currently on the land. The commercial cropland was assumed to have had zero tillage management in the past when determining the equilibrium carbon stock at time zero.

The amount of carbon already present in the soil is the same as for the preceding calculation for the project-specific commercial cropland carbon stock.

SOC = 50 t C/ha • 0.82 • 1.10 • 1.0[14]

= 45.10 t C/ha

The carbon stocks on the project site are the sum of the living biomass and soil carbon pools, plus the carbon stock at time zero. The average carbon stocks and the associated change in carbon stocks for the project site are given in table E.18. Table E.19 shows the increasing carbon stocks per hectare and the total carbon stocks for the project activity.

The GHG Reduction

Project-Specific Procedure

The total carbon stocks and GHG removals per hectare attributed to the Nipawin Afforestation Project are calculated in tables E.18 and E.19. The GHG reduction is calculated as the difference between the project activity's GHG removals (i.e., Nipawin Afforestation Project) and the baseline GHG

TABLE E.16 Finding the Hectares by which to Weight the Baseline Carbon Stocks

	COMMERCIAL CROPLAND	PASTURELAND	HYBRID POPLAR FOREST
Percent of geographic area	88.5	11	.5
Project site hectares	26,550	3,300	150

TABLE E.17 Total Baseline Carbon Stock for the Performance Standard

YEAR	COMMERCIAL CROPLAND t C/ha	PASTURELAND t C/ha	HYBRID POPLAR FOREST t C/ha	TOTAL CARBON STOCKS t C
1	45.1	55	45.95	1,385,798
2	45.1	55	48.21	1,386,136.5
3	45.1	55	52.33	1,386,754.5
4	45.1	55	58.17	1,387,630.5
5	45.1	55	65.36	1,388,709
6	45.1	55	73.49	1,389,928.5
7	45.1	55	82.16	1,391,229
8	45.1	55	90.99	1,392,553.5
9	45.1	55	99.76	1,393,869
10	45.1	55	108.25	1,395,142.5
11	45.1	55	116.33	1,396,354.5
12	45.1	55	123.91	1,397,491.5
13	45.1	55	130.95	1,398,548
14	45.1	55	137.41	1,399,516.5
15	45.1	55	143.32	1,400,403
16	45.1	55	148.69	1,401,208.5
17	45.1	55	153.55	1,401,937.5
18	45.1	55	157.93	1,402,594.5
19	45.1	55	161.88	1,403,187
20	45.1	55	165.42	1,403,718
21	45.1	55	165.42	1,403,718

BOX E.2 Calculating the Aboveground Biomass for Year 5 of the Nipawin Afforestation Project

Aboveground biomass

= sum of carbon sequestered in aboveground biomass by five 1500-ha parcels that were afforested

= (17.34 t C ha/yr • 1500 ha) + (10.74 t C/ha/yr • 1500 ha) + (5.52 t C/ha/yr • 1500 ha) + (2.00 t C/ha/yr • 1500 ha) + (0.31 t C/ha/yr • 1500 ha)

= 53,865 t C in year 5

The aboveground biomass per hectare in year 5 is found by dividing 53,865 t C by 30,000.

TABLE E.18 Living Biomass and Soil Carbon Stocks, Change in Carbon Stocks, and GHG Removals for the Nipawin Afforestation Project

YEAR	CUMULATIVE INCREASE IN ABOVEGROUND CARBON STOCKS	CUMULATIVE INCREASE IN BELOWGROUND CARBON STOCKS	CUMULATIVE INCREASE IN SOIL CARBON STOCKS	TOTAL CUMULATIVE INCREASE IN CARBON STOCKS	CHANGE IN CARBON STOCKS	GHG REMOVALS
	t C/ha	t C/ha	t C/ha	t C/ha	t C/ha	t CO_2/ha
1	0.02	0	0.03	0.05	0.05	0.18
2	0.12	0.01	0.08	0.21	0.16	0.59
3	0.39	0.02	0.15	0.56	0.35	1.28
4	0.93	0.03	0.25	1.21	0.65	2.38
5	1.8	0.04	0.38	2.22	1.01	3.70
6	3.04	0.06	0.53	3.63	1.41	5.17
7	4.69	0.08	0.7	5.47	1.84	6.75
8	6.75	0.1	0.9	7.75	2.28	8.36
9	9.21	0.12	1.13	10.46	2.71	9.94
10	12.08	0.14	1.38	13.6	3.14	11.51
11	15.32	0.16	1.65	17.13	3.53	12.94
12	18.91	0.18	1.95	21.04	3.91	14.34
13	22.82	0.2	2.28	25.3	4.26	15.62
14	27.03	0.23	2.63	29.89	4.59	16.83
15	31.51	0.25	3	34.76	4.87	17.86
16	36.23	0.27	3.4	39.9	5.14	18.85
17	41.17	0.29	3.83	45.29	5.39	19.76
18	46.3	0.31	4.28	50.89	5.6	20.53
19	51.6	0.34	4.75	56.69	5.8	21.27
20	57.05	0.36	5.25	62.66	5.97	21.89
21	57.05	0.36	5.25	62.66	0	0

BOX E.3 Calculating the Soil Carbon for Year 5 of the Nipawin Afforestation Project

Soil carbon

= sum of carbon sequestered in the soil by three 1500-ha parcels that were afforested

= (2.5 t C ha/yr • 1500 ha) + (2.0 t C/ha/yr • 1500 ha) + (1.5 t C/ha/yr • 1500 ha) + (1.0 t C ha/yr • 1500 ha) + (0.5 t C ha/yr • 1500 ha)

= 3750 t C/yr + 3000 t C/yr + 2250 t C/yr + 1500 t C/yr + 750 t C/yr

= 11,250 t C in year 5

The soil carbon per hectare for year 5 is found by dividing 11,250 t C by 30,000.

EXAMPLE: Nipawin Afforestation Project

TABLE E.19 Total Carbon Stocks for Project Activity

YEAR	CARBON STOCKS PER HECTARE t C/ha	TOTAL CARBON STOCK t C
0	45.1	1,353,000
1	45.15	1,354,500
2	45.31	1,359,300
3	45.66	1,369,800
4	46.31	1,389,300
5	47.32	1,419,600
6	48.73	1,461,900
7	50.57	1,517,100
8	52.85	1,585,500
9	55.56	1,666,800
10	58.7	1,761,000
11	62.23	1,866,900
12	66.14	1,984,200
13	70.4	2,112,000
14	74.99	2,249,700
15	79.86	2,395,800
16	85	2,550,000
17	90.39	2,711,700
18	95.99	2,879,700
19	101.79	3,053,700
20	107.76	3,232,800
21	107.76	3,232,800

TABLE E.20 GHG Reductions Using the Project-Specific Procedure

YEAR	TOTAL GHG REDUCTION t CO_2
1	5,500
2	17,600
3	38,500
4	71,500
5	111,100
6	155,100
7	202,400
8	250,800
9	298,100
10	345,400
11	388,300
12	430,100
13	468,600
14	504,900
15	535,700
16	565,400
17	592,900
18	616,000
19	638,000
20	656,700
21	0

removals corresponding to the baseline scenario for commercial cropland in chapter 6. No secondary effects are included.

The GHG removals from the commercial cropland baseline scenario are equal to zero. Therefore, the GHG reductions from the Nipawin Afforestation Project are equal to the GHG removals from the project activity (see table E.18) multiplied by 30,000 hectares. Table E.20 shows the GHG reductions.

Performance Standard Procedure
The total carbon stocks and GHG removals attributed to the Nipawin Afforestation Project were calculated earlier. The GHG reduction is calculated as the difference between the project activity's GHG removals (i.e., Nipawin Afforestation

TABLE E.21 GHG Removals for the Performance Standard and the Nipawin Afforestation Project, and the GHG Reduction

YEAR	GHG REMOVALS: PERFORMANCE STANDARD	GHG REMOVALS: NIPAWIN AFFORESTATION PROJECT	GHG REDUCTION
	t CO_2/ha	t CO_2/ha	t CO_2/ha
1	0.00059	0.18	0.1827
2	0.00158	0.59	0.5851
3	0.00287	1.28	1.2805
4	0.00407	2.38	2.3793
5	0.00501	3.70	3.6983
6	0.00567	5.17	5.1643
7	0.00604	6.75	6.7406
8	0.00615	8.36	8.3538
9	0.00611	9.94	9.9306
10	0.00592	11.51	11.5074
11	0.00563	12.94	12.9377
12	0.00528	14.34	14.3314
13	0.00491	15.62	15.6151
14	0.00450	16.83	16.8255
15	0.00412	17.86	17.8525
16	0.00374	18.85	18.8429
17	0.00339	19.76	19.7599
18	0.00305	20.53	20.5303
19	0.00275	21.27	21.2639
20	0.00247	21.89	21.8875
21	0	0	0

80

Chapter 11: Carbon Reversibility Management Plan

The two risks to hybrid poplar forests are fire and insects. The carbon reversibility management plan therefore uses direct suppression, replacement, insurance, and buffering strategies to mitigate carbon reversal.

1. **Direct suppression:** The risk of fire is relatively low for hybrid poplar plantations, and the Nipawin Afforestation Project is located in intensively managed agricultural areas, thus reducing even further the risk of fire. Poplar is also relatively inflammable compared with the spruce located in the boreal forest farther north. Because hybrid poplar is a species new to this geographic area, little information is available about the probability of insect infestation, but it will be carefully monitored throughout the project's life.

2. **Replacement:** Any carbon that is lost because of fire or insect damage will be replaced with additional GHG reductions bought from the market.

3. **Insurance:** Insurance companies offer plantations insurance for losses from insect or fire damage. The insurance payments from any carbon loss from fire or insects will be used to buy replacement GHG reductions from the market.

4. **Buffering:** As an additional precaution, a proportion of the GHG reductions from the Nipawin Afforestation Project will be withheld to help cover any carbon reversals caused by unpredictable natural disturbances such as insects and fire. The risk of a future loss of carbon sequestered as a result of insects and fire is 15 percent in old natural forest in Saskatchewan (Lemprière et al. 2002). For afforestation, the risk of loss is smaller because the plantations are located in intensively

Project) and the performance standard. The performance standard stringency level selected was the weighted average. Table E.21 gives the performance standard GHG removals, the Nipawin Afforestation project GHG removals, and the GHG reduction. No secondary effects are included.

Figure E.2 shows that the difference in the GHG removals between the performance standard and the project activity is small initially, but increases each year. By year 20, the difference in the GHG removals between the performance standard and the project activity is almost 22 t CO_2/ha. Using the baseline scenario (project-specific procedure) the baseline GHG removals would be very similar. The total GHG reductions for the 30,000 hectares in tonnes carbon from the GHG project are shown in table E.22.

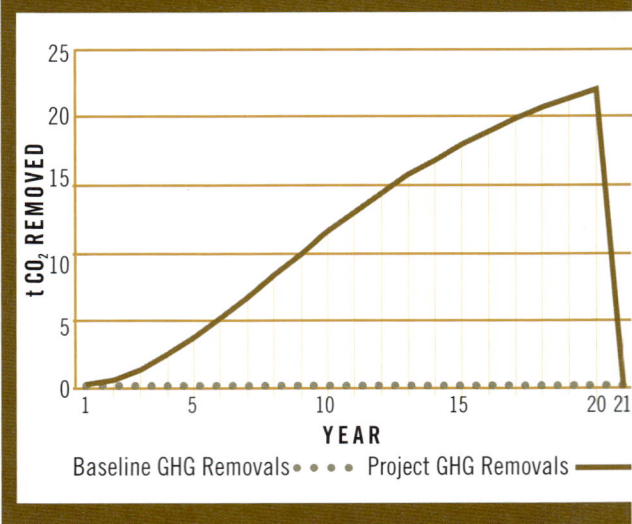

FIGURE E.2 The Difference between the Performance Standard and Project Activity's GHG Removals

Baseline GHG Removals • • • Project GHG Removals ——

EXAMPLE: Nipawin Afforestation Project

TABLE E.22 GHG Reductions Using the Performance Standard Procedure

YEAR	TOTAL GHG REDUCTION t CO_2
1	5,482.23
2	17,552.75
3	38,413.86
4	71,377.89
5	110,949.67
6	154,930.01
7	202,218.72
8	250,615.38
9	297,916.63
10	345,222.48
11	388,131.06
12	429,941.51
13	468,452.80
14	504,764.93
15	535,576.43
16	565,287.93
17	592,798.17
18	615,908.42
19	637,917.41
20	656,625.98
21	0

NOTES

[1] These soil zones were identified as relevant when identifying the geographic area, see section 5.3.

[2] Because the Census Agricultural Regions are quite large, the regions determined to be Dark Gray and/or Black are a best estimate.

[3] The final area used to define the performance standard might still be a subsection of this broad area, as the final list of baseline candidates may represent a subsection of this total area, e.g., natural pasture lands need to be excluded from this area.

[4] As used in this document, agricultural land refers to land classified as "Total Area of Farms" in the Census Agricultural Regions in Alberta, Saskatchewan, and Manitoba (Statistics Canada 1996, 2001).

[5] Internal WRI source, not published material.

[6] Lemprière et al. 2002.

[7] The F_{LU}, F_{MG}, and F_I default values can be found in the IPCC (2003).

[8] The default values can be found in the IPCC (2003).

[9] Personal communication with Al-Pac personnel.

[10] Tier 1 IPCC assumption.

[11] The F_{LU}, F_{MG}, and F_I default values can be found in the IPCC (2003).

[12] The default values can be found in the IPCC (2003).

[13] The default values can be found in the IPCC (2003).

[14] The default values can be found in the IPCC (2003).

managed agricultural land, so a 5 percent buffer would be reasonable. Accordingly, 95 percent of the GHG reductions from the Nipawin Afforestation Project will be traded on the market, and 5 percent will be withheld.

To estimate the magnitude of the primary effect, above- and belowground living biomass and soil carbon will be monitored as they increase as a result of the project activity. Dead biomass does not accumulate in significant amounts and so will not be monitored. The monitoring follows the IPCC's *Good Practice Guidance for LULUCF* (2003) and the Offset System Quantification Protocol for Afforestation Projects (Graham 2006).

In addition, the National Forest Information System (Natural Resources Canada 2006) will be monitored for the new afforestation developments in the geographic area. If the afforestation becomes substantial, the baseline GHG removals will need to be re-estimated or a land-use trend factor will have to be added.

Annex A

Life-Cycle Assessments and Upstream and Downstream Effects

Table A.1 gives an example of a full life-cycle assessment of a willow reforestation project in central New York State (adapted from Heller, Keoleian, and Volk 2003). *A life-cycle assessment* quantifies the possible secondary effects by looking at the material and energy inputs and outputs throughout a product's life, from raw material acquisition through production, use, and disposal. The assessment does not consider market responses, and the primary effect's GHG removals pertain to the carbon sequestered in above- and belowground biomass.

This example shows that
- Carbon sequestered in vegetation can outweigh the GHG emissions from secondary effect sources.

- Many secondary effects would probably be considered insignificant compared with the primary effect.

Figure A.1 shows how the GHG effects from the life-cycle assessment may be used to define the GHG assessment boundary. Even though this boundary does not include all the GHG effects from a life-cycle assessment, it is important to consider these secondary effects initially to ensure a complete accounting of them.

TABLE A.1 Magnitude of Primary and Secondary Effects Based on a Life-Cycle Assessment

	CO_2	OTHER GHGS	GHG TOTAL
	Mg CO_2 eq/ha		
Sources of Potential Secondary Effects			
GHG emissions			
• Diesel fuel	+3.12[a]	+0.06	+3.2
• Agricultural inputs[b]	+3.0	+0.4	+3.4
• N_2O from applied N for fertilizer		+4.0(±3.2)[c]	+4.0
• N_2O from leaf litter		+7.3(±5.8)[c]	+7.3
Primary Effect Sinks			
C sequestration			
• Belowground biomass	−14.1		−14.1
• Soil carbon[d]	0		0
• Harvestable biomass	−499.2		−499.2

Notes:
[a] Positive values indicate additions (releases) to the atmosphere.

[b] Includes fertilizer and herbicide manufacturing and transport, machinery manufacturing, and nursery operations.

[c] Bracketed numbers represent the N_2O emission range presented by the IPCC (1997) estimate.

[d] Soil carbon was assumed to be zero owing to uncertainty in measurements and variability in the study area.

FIGURE A.1 Diagram of Possible GHG Effects in the Life-Cycle Assessment and the GHG Assessment Boundary for the Project Activity

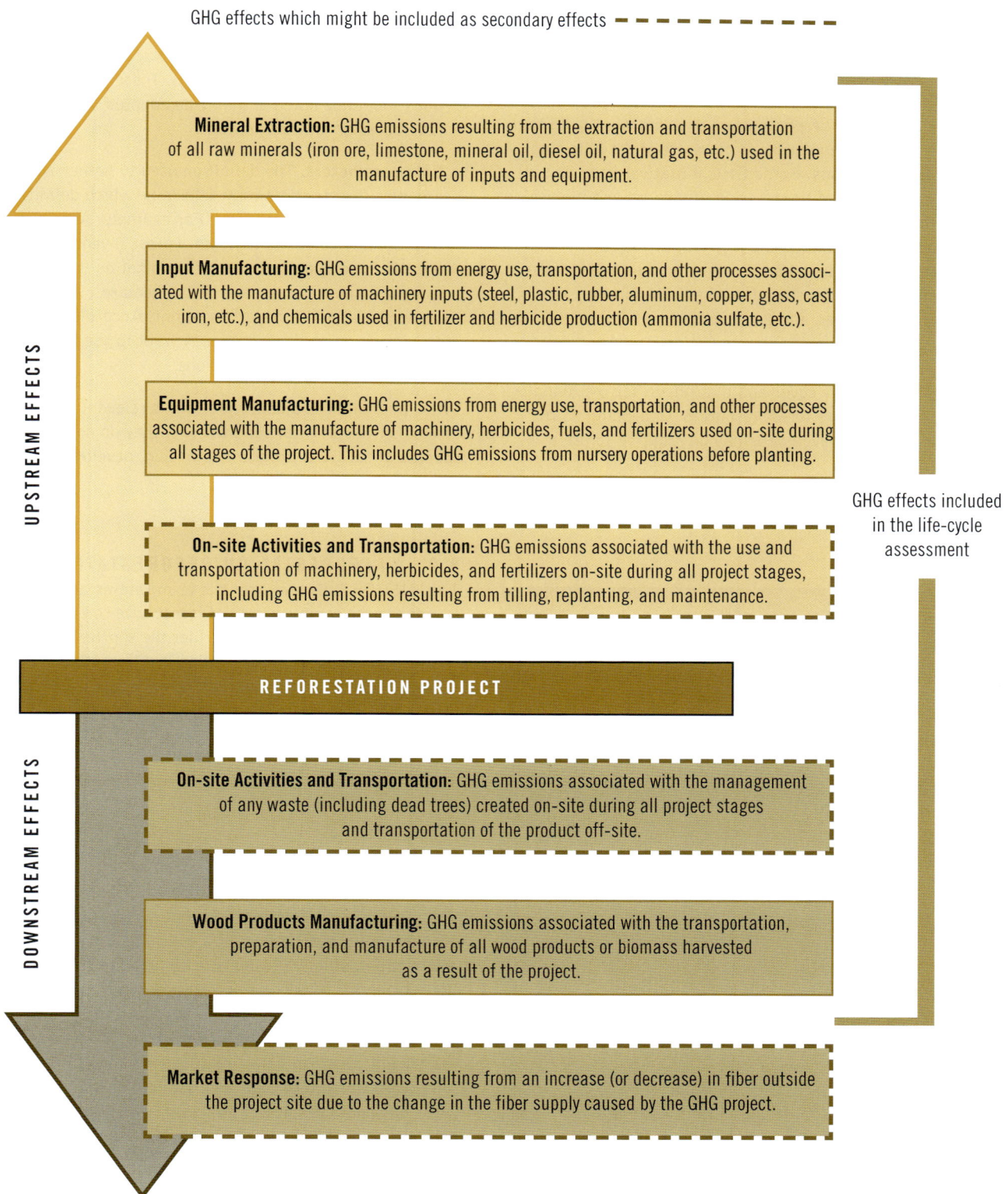

GHG effects which might be included as secondary effects – – – – – – – – – –

UPSTREAM EFFECTS

Mineral Extraction: GHG emissions resulting from the extraction and transportation of all raw minerals (iron ore, limestone, mineral oil, diesel oil, natural gas, etc.) used in the manufacture of inputs and equipment.

Input Manufacturing: GHG emissions from energy use, transportation, and other processes associated with the manufacture of machinery inputs (steel, plastic, rubber, aluminum, copper, glass, cast iron, etc.), and chemicals used in fertilizer and herbicide production (ammonia sulfate, etc.).

Equipment Manufacturing: GHG emissions from energy use, transportation, and other processes associated with the manufacture of machinery, herbicides, fuels, and fertilizers used on-site during all stages of the project. This includes GHG emissions from nursery operations before planting.

On-site Activities and Transportation: GHG emissions associated with the use and transportation of machinery, herbicides, and fertilizers on-site during all project stages, including GHG emissions resulting from tilling, replanting, and maintenance.

GHG effects included in the life-cycle assessment

REFORESTATION PROJECT

DOWNSTREAM EFFECTS

On-site Activities and Transportation: GHG emissions associated with the management of any waste (including dead trees) created on-site during all project stages and transportation of the product off-site.

Wood Products Manufacturing: GHG emissions associated with the transportation, preparation, and manufacture of all wood products or biomass harvested as a result of the project.

Market Response: GHG emissions resulting from an increase (or decrease) in fiber outside the project site due to the change in the fiber supply caused by the GHG project.

Annex B

Forest and Project Activity Definitions

B.1   FOREST DEFINITIONS

Because of differences among program definitions, data collection parameters, national regulations, and environmental conditions, the LULUCF Guidance does not offer an exact definition of forest. Therefore, when deciding what definition to use, project developers should be aware of the following:

- **Program regulations.** Some GHG initiatives have a specific definition of 'forest,' and others use less precise definitions, leaving room for interpretation and regional conditions. For example, the New South Wales Greenhouse Gas Abatement Scheme defines a forest as land with a minimum size of 0.2 hectares in area, 20 percent crown cover, and 2 meters of tree height. In contrast, the California Climate Action Registry defines a forest as land that supports, or can support, at least 10 percent tree canopy cover and that allows for the management of one or more forest resources. Depending on the program, project developers may be required to use a specific definition.

- **National regulations.** By the end of 2006, parties to the Kyoto Protocol must decide on their definitions of forest, choosing a minimum area (0.05 to 1.0 ha), a minimum crown cover at maturity (10 to 30%), and a minimum tree height at maturity (2 to 5 m). These national definitions should be noted and may be adopted by project developers or future programs.

- **Data parameters.** The definition used to gather and organize forestry data may determine which definition a project developer should use. For example, Australia has traditionally compiled National Forest Inventory data for the UN Food and Agriculture Organization (UN FAO) using minimum parameters of 1.0 hectares, 20 percent crown cover, and 2 meters in height. Accordingly, project developers using these data may want to use the same definition to avoid complications.

When possible, use a definition that is relevant to regional or local environmental conditions. If they do not have a specific or set GHG program definition, developers should use the UN FAO definitions.

B.2   AFFORESTATION/REFORESTATION

The LULUCF Guidance uses the term reforestation broadly to refer to the establishment of forest cover on cleared land that was previously forest (in either the shorter or longer term). In most cases, reforestation is differentiated from afforestation by the time period in which forest last existed on the land. For reforestation, the time period is usually shorter than for afforestation, for example, reforested land may have been in forest just five years previously, whereas afforestation is the creation of forest on land considered not to have been forest for a longer term (e.g., fifty years).

The reason that the LULUCF Guidance refers to reforestation rather than afforestation is that even though the GHG accounting for both activities is the same, determining afforestation efficiently and consistently can be difficult. GHG programs allowing reforestation projects may use a set base year (e.g., 1990) or may determine a minimum period during which the land must have been out of forest (e.g., ten years), which can be shown relatively simply using remotely sensed data. Showing that land has not been under forest cover for longer periods of time may be more difficult. The LULUCF Guidance is not, however, trying to specifically define either afforestation or reforestation.

Annex C

Additional QA/QC Guidance

Because forest measurements are sampled at a low intensity, monitoring methodologies should have adequate controls in place to make sure that errors are not amplified within the estimation process. Project developers should document and, in some instances, justify the selected sampling approach, field measurement procedures, training procedures, internal auditing system, data entry and analysis, and data-archiving procedures.

The following should be included in the QA/QC section of the monitoring plan, which, however, should not replace a detailed description of the monitoring methodology in general.

C.1   DOCUMENT CORE ELEMENTS OF A SAMPLING/MONITORING APPROACH

Project developers should provide documentation and justification of the carbon pool sampling approach they use, including
- A description of the size and variability of the resource and the factors determining the approach to achieving adequate levels of accuracy, precision, and cost-effectiveness.

- An overview of the sampling approach with the specific
  - Target precision levels.

  - Stratification of the resource.

  - Number of plots per strata plus a description of when any deviations may occur.

  - Types of plots used (size, temporary or permanent, fixed or variable area, circular or quadratic).

  - Sampling schedule or duration between measurements.

  - Any other information relevant to the quality and outcomes of the selected approach.

- A description of how the approach adequately allows for second- and third-party auditing or plot remeasurement.

- A reference to the field measurement procedures to be applied, or the standard operating procedures (SOPs).

C.2   DOCUMENT FIELD MEASUREMENT PROCEDURES/STANDARD OPERATING PROCEDURES

Project developers should describe in detail the procedures to be used for the field measurements. These procedures should spell out clearly and concisely each aspect of the field measurement practices. Project developers should also provide
- A description of the mode of field data capture (i.e., paper or digital, type of digital capture technology).

- A description of the internal control processes in place for each field measurement activity that minimizes the risk of collecting incorrect data.

C.3   DOCUMENT PERSONNEL AND TRAINING

Project developers should describe all parties taking field measurements and their training, including
- The names and roles of all parties taking field measurements.

- The status of each crew member (i.e., whether they are students, internal part-time measurement crew members, internal full-time measurement crew members, or external measurement contractors).

- A description of how training is undertaken or minimum knowledge is taught to all involved parties, including how crew members have been taught the field measurement procedures/SOPs.

- A description of how new or replacement crew members are trained.

- Any stipulations of task or role restrictions for new or minimally trained crew members.

- The names and a description of those responsible for second-party auditing or remeasurement, including their level of knowledge, training, and experience.

C.4   INTERNAL AUDITING/ PLOT REMEASUREMENT

Project developers should describe the process in place for second-party auditing or plot remeasurement, including
- A stipulation and justification of the target level of remeasurement as a percentage of the total number of annual plots, as well as a description of such aspects as the crew's experience.

- A description of the minimum standards of acceptable plot measurement accuracy and the responses to various outcomes.

- A description of how plot relocation and remeasurement have been ensured, plus a description of the storage of GPS plot location files and how exact plot centers (or equivalent) have been marked so that they can be found after either
  - A minimum of twelve months from the time of plot establishment for a GHG program using a random auditing approach.

  - A minimum of five years from the time of plot establishment.

- A description of the system used to store, analyze, and report auditing information and outcomes.

85

• A description of how field measurement accuracy outcomes have been incorporated into the estimation process.

C.5   DATA ENTRY AND ANALYSIS

Project developers should document the process and procedures for ensuring reliable and accurate data entry and analysis, including

• The names, experience, and specific roles of all parties in the data entry and analysis.

• A description of the upload, database, and file write access restrictions contributing to the data's reliability.

• A description of the internal control processes in place for each data entry and analysis activity minimizing the risk of incorrect data entry and/or use.

C.6   DATA MAINTENANCE AND STORAGE/ARCHIVING

Project developers should document how the data are to be maintained in the long term, including
• A description of how plot sheets are stored (if paper is used).

• A description of how the original field data files are being stored.

• A description of the procedures in place for the updating or conversion of files from older to current digital file formats.

• A description of the actions taken to minimize the risk of database corruption, including backups and database and file write access restrictions.

• A description of any archiving procedures or protocols.

NOTES
[1] The guidance in annex C was developed by Penny Baalman from Del Norske Veritas (DNV).

Glossary

Most of the definitions for the following terms are the same as those in the Project Protocol, although some have been altered and others not relevant to LULUCF projects have been omitted.

Additionality
A criterion often applied to GHG projects, stipulating that project-based GHG reductions should be quantified only if the project activity "would not have happened anyway," that is, that the project activity (or the same land-use or management practices it employs) would not have been implemented in its baseline scenario and/or that the project activity GHG removals are greater than the baseline GHG removals.

Afforestation
The creation of forest on land that is considered to not have been forest previously, at least for a long time (e.g., fifty years). For additional information on the distinction between afforestation and reforestation and its use in this guidance, see annex B.

Allowances
The basic tradable commodity in GHG emission–trading systems. Allowances grant their holder the right to emit a specific quantity of pollution once (e.g., one tonne of CO_2eq). The total quantity of allowances issued by regulators dictates the total quantity of emissions possible under the system. At the end of each compliance period, each regulated entity must surrender sufficient allowances to cover its GHG emissions during that period.

Barriers
Any factor or consideration that would (significantly) discourage a decision to try to implement the project activity or its baseline candidates.

Baseline Candidates
The alternative land uses or management practices on lands located in a specific geographic area and during a given temporal range.

Baseline Parameter
Any parameter whose value or status can be monitored in order to validate assumptions about baseline removal estimates or to help estimate baseline removals.

Baseline Procedures
Methods used to estimate baseline emissions and removals. The Project Protocol offers two optional procedures: the project-specific procedure and the performance standard procedure.

Baseline Scenario
A hypothetical description of what most likely would have occurred in the absence of any mitigation of climate change.

Benefits
The benefits expected to accrue to decision makers for the activities in each baseline scenario alternative, excluding all potential benefits resulting from GHG reductions.

Carbon Dioxide Equivalent (CO_2eq)
The universal unit of measurement used to indicate the global-warming potential of greenhouse gases. It is used to evaluate the impacts of releasing (or avoiding the release of) different greenhouse gases.

Carbon Pool
A reservoir or a system with the capacity to accumulate or release carbon. Examples of carbon pools are living biomass, dead organic mater and soils. The units used are mass (e.g., t C).

Carbon Stock
The absolute quantity of carbon held in a carbon pool(s) at a specified time (see GHG Sink).

Common Practice
The predominant land use(s) or management practice(s) undertaken in a particular region or sector.

Deforestation
The removal of forest cover to the extent that the land is transformed from forest to nonforest land. Although it does not explicitly discuss deforestation projects, this guidance may be applied to them.

Dynamic Baseline Removals
Baseline removal estimates that change over the valid time length of the baseline scenario.

Equilibrium Carbon Stocks
The carbon stored at time zero of the project's implementation; when there has been no change in land use or management or other events that might affect any of the carbon pools, and the carbon storage is at a steady state.

Emission Factor
A factor relating GHG emissions to a level of activity or a certain quantity of inputs or products or services (e.g., tonnes of fuel consumed or units of a product). For example, an electricity emission factor is commonly expressed as t CO_2eq/megawatt-hour.

87

Forest	For a definition of forest, see annex B. In the absence of GHG program or other specific definitions, project developers should use the following definition provided by the UN FAO: Forest includes natural forests and forest plantations. It is used to refer to land with a tree canopy cover of more than 10 percent and area of more than 0.5 ha. Forests are determined both by the presence of trees and the absence of other predominant land uses. The trees should be able to reach a minimum height of 5 m. Young stands that have not yet but are expected to reach a crown density of 10 percent and tree height of 5 m are included under forest, as are temporarily unstocked areas. The term includes forests used for purposes of production, protection, multiple-use or conservation (i.e., forest in national parks, nature reserves and other protected areas), as well as forest stands on agricultural lands (e.g., windbreaks and shelterbelts of trees with a width of more than 20 m), and rubberwood plantations and cork oak stands. The term specifically excludes stands of trees established primarily for agricultural production, for example fruit tree plantations. It also excludes trees planted in agroforestry systems.
Geographic Area	A physical area that helps define the final list of baseline candidates. The area can be defined by a number of factors, including biophysical characteristics, sociocultural, economic, or legal factors; and/or the availability of necessary physical infrastructure.
GHG Assessment Boundary	A boundary encompassing all primary effects and significant secondary effects associated with the GHG project. If the GHG project has more than one activity, the primary and significant secondary effects from all the activities are included in the GHG assessment boundary.
GHG Emissions	GHGs released into the atmosphere.
GHG Program	A generic term for (1) any voluntary or mandatory, government or nongovernment initiative, system, or program that registers, certifies, or regulates GHG emissions; or (2) any authorities responsible for developing or administering such initiatives, systems, or programs.
GHG Project	A specific activity or set of activities intended to reduce GHG emissions, increase the storage of carbon, or enhance GHG removals from the atmosphere. A GHG project may be a stand-alone project or a component of a larger non-GHG project.
GHG Protocol Initiative (GHG Protocol)	A multistakeholder partnership of businesses, nongovernmental organizations, governments, academics, and others convened by the World Business Council for Sustainable Development and the World Resources Institute to design and develop internationally accepted GHG accounting and reporting standards and/or protocols and to promote their broad adoption.
GHG Reductions	A decrease in GHG emissions or an increase in the removals and storage of GHGs from the atmosphere, relative to baseline emissions/removals. Primary effects result in GHG reductions, as do some secondary effects. A project activity's total GHG reductions are quantified as the sum of its associated primary effect(s) and any significant secondary effects (which may be decreases or countervailing increases in GHG emissions). A GHG project's total GHG reductions are quantified as the sum of the GHG reductions from each project activity.
GHG Removal	The storage of carbon dioxide between two points of time. For LULUCF projects, GHG removals are found by first finding the change in carbon stocks between two time periods, and multiplying the carbon stock change by $\frac{44}{12}$ t CO_2/ t C. This definition differs slightly from the use of "GHG removal" in the Project Protocol, where it refers only to biological sequestration rather than the comparison of two points in time, as in this document.
GHG Sink	Any process or mechanism that removes from the atmosphere a greenhouse gas, an aerosol, or a precursor of a greenhouse gas. A carbon pool (reservoir) can be a sink for atmospheric carbon if, during a given time interval, more carbon is flowing into it than is flowing out.
GHG Source	Any process that releases GHG emissions into the atmosphere. The five general GHG source categories are • Combustion emissions from generating grid-connected electricity. • Combustion emissions from generating non-grid-connected electricity or energy. • Process emissions from industrial activities. • Fugitive emissions. • Waste emissions.

Greenhouse Gases (GHGs)	Greenhouse gases are gases that absorb and emit radiation at specific wavelengths within the spectrum of infrared radiation emitted by the earth's surface, the atmosphere, and clouds. The six main GHGs whose emissions are caused by humans are carbon dioxide (CO_2), methane (CH_4), nitrous oxide (N_2O), hydrofluorocarbons (HFCs), perfluorocarbons (PFCs), and sulfur hexafluoride (SF_6).
Land Use	All the arrangements, activities, and inputs undertaken in a certain land cover type (a set of human actions) or the social and economic purposes for which land is managed (e.g., grazing, timber extraction, conservation).
Legal Requirements	Any mandatory laws or regulations that directly or indirectly affect the GHG emissions or removals associated with a project activity or its baseline candidates and that require technical, performance, or management actions. Legal requirements may involve using a specific technology (e.g., gas turbines instead of diesel generators), meeting a certain standard of performance (e.g., fuel efficiency standards for vehicles), or managing operations according to certain criteria or practices (e.g., forest management practices).
Management Practice	An action or set of actions that affect the land, the carbon stocks in pools associated with it, or the exchange of greenhouse gases with the atmosphere.
Market Response	The response of alternative providers or users of an input or product to a change in the market supply or demand caused by the project activity.
One-Time Effects	Secondary effects related to a project activity's construction, installation, and establishment or its decommission and termination.
Performance Metric	A rate that relates the GHG removals to the size of different baseline candidates for a specific time period, t. Performance metrics are used in developing performance standards. For LULUCF projects, the performance metric is $$\frac{\text{GHG Removals}_t}{\text{Unit of land area}}$$
Performance Standard	A GHG removal rate used to determine baseline GHG removals for a particular type of project activity. A performance standard may be used to estimate baseline GHG removals for any number of similar project activities in the same geographic area.
Performance Standard Procedure	A baseline procedure that estimates baseline GHG removals using a GHG removal rate derived from a numerical analysis of the GHG removal rates of all baseline candidates. A performance standard is sometimes referred to as a multi-project baseline or benchmark because it can be used to estimate baseline GHG removals for multiple project activities of the same type.
Permanence	The longevity of a carbon pool and the stability of its stocks within its management and disturbance environment.
Primary Effect	The intended change caused by a project activity in GHG emissions, removals, or storage associated with a GHG source or sink. For LULUCF project activities, this includes all biological carbon stock changes caused by the project activity on the project site. Each project activity generally has only one primary effect. The primary effect must be defined relative to baseline GHG removals.
Project	See GHG Project.
Project Activity	A specific action or intervention targeted at changing GHG emissions, removals, or storage. It may include modifications of or alterations to existing production, process, consumption, service, or management systems, as well as the introduction of new systems.
Project Developer	A person, company, or organization developing a GHG project.
Project-Specific Procedure	A baseline procedure that estimates baseline GHG removals by identifying a baseline scenario specific to the proposed project activity.
Reforestation	The reestablishment of forest cover on cleared land that previously was forest. For additional information on the distinction between afforestation and reforestation and its use in this guidance, see annex B.

89

Secondary Effect

An unintended change caused by a project activity in GHG emissions, removals, or storage associated with a GHG source or sink. For LULUCF project activities, all secondary effects are nonbiological GHG changes caused by the project activity and any biological carbon stock changes that occur off the project site, for example, owing to a market response. Secondary effects are typically small relative to a project activity's primary effect, although sometimes they may undermine or negate the primary effect. Secondary effects are classified as

- One-time effects, changes in GHG emissions associated with the construction/installation/ establishment or decommission/termination of the project activity
- Upstream and downstream effects, changes in GHG emissions associated with inputs to the project activity (upstream) or products from the project activity (downstream), relative to the baseline emissions. Upstream and downstream effects may provoke market responses to the changes in supply and/or demand for project activity inputs or products.

Sequestration

The uptake and storage of CO_2, which can be sequestered by plants or in underground or deep-sea reservoirs.

Stringency Level

A GHG removal rate that is more restrictive than the average GHG removal rate of all baseline candidates. Stringency levels may be specified as a GHG removal rate corresponding to a certain percentile (better than the 50th percentile) or to the baseline candidate with the highest removal rate. Stringency levels are defined in the course of developing a performance standard.

Temporal Range

A contiguous time period that helps define the final list of baseline candidates. The temporal range can be defined by a number of factors, such as the dominance of a single practice for an extended period of time, the diversity of options in a sector or region, and/or a discrete change in an area's or a region's policy, technology, practice, or resource.

Time-Based Performance Standard

A performance standard defined as the rate of GHG removals for a baseline candidate's unit of time and land area.

Upstream/Downstream Effects

Secondary effects associated with the inputs used (upstream) or the products produced (downstream) by a project activity.

Wood Products

Products produced from harvested fiber, including fuel wood, logs, and the processed products such as sawn timber, plywood, wood pulp, and paper.

References

The Additional Resources listed after the references provide sources of information that discuss in more detail the various concepts described in the relevant chapters. The Project Protocol also lists other references and sources of information for the various chapters.

CHAPTER 3: DEFINING THE GHG ASSESSMENT BOUNDARY

Aukland, L., P. Moura Coasta, and S. Brown. 2002. A Conceptual Framework and Its Application for Addressing Leakage: The Case of Avoided Deforestation. *Climate Policy* 94:1–15.

Intergovernmental Panel on Climate Change (IPCC). 1997. *Revised 1996 IPCC Guidelines for National Greenhouse Gas Inventories,* edited by J.T. Houghton et al. Bracknell: U.K. Meteorological Office, IPCC/OECD/IEA.

Murray, B.C., B.A. McCarl, and H. Lee. 2004. Estimating Leakage from Forest Carbon Sequestration Programs. *Land Economics,* February.

Additional Resources

Adams, Darius M.; Alig, Ralph J.; Callaway, J.M.; McCarl, Bruce A.; Winnett, Steven M. 1996. The forest and agricultural sector optimization model (FASOM): Model Structure and Policy Applications. Research Paper PNW-RP-495. Portland, OR: U.S. Department of Agriculture, Forest Service, Pacific Northwest Research Station. 60 p.

Center for International Forestry Research (CIFOR). 2001. A Shared Research Agenda for Land-Use, Land Use Change, Forestry and the Clean Development Mechanism. Bogor: CIFOR.

Chomitz, K. 2002. Baseline, Leakage and Measurement Issues: How Do Forestry and Energy Projects Compare. *Climate Policy* 2:35–49.

Intergovernmental Panel on Climate Change (IPCC). 2003. *Good Practice Guidance for Land Use, Land-Use Change and Forestry,* edited by J. Penman et al. Hayama: Institute for Global Environmental Strategies.

International Panel on Climate Change (IPCC). 2006. *2006 IPCC Guidelines for National Greenhouse Gas Inventories,* edited by S. Eggleston et al. Hayama: Institute for Global Environmental Strategies.

Merry, F.D., and D.R. Carter. 2001. Factors Affecting Bolivian Mahogany Exports with Policy Implications for the Forest Sector. *Forest Policy and Economics* 2:281–90.

Niles, J.O., S. Brown, J. Pretty, A.S. Ball, and J. Fay. 2002. *Potential Carbon Mitigation and Income in Developing Countries from Changes in Use and Management of Agricultural and Forest Lands.* London: Royal Society.

Schwarze, R., J.O. Niles, and J. Olander. 2002. *Understanding and Managing Leakage in Forest-Based Greenhouse-Gas-Mitigation Projects.* London: Royal Society.

Sohngen, B., and S. Brown. 2004. *Measuring Leakage from Carbon Projects in Open Economies: A Stop Timber Harvesting Project in Bolivia as a Case Study.* Mississauga: NRC Canada.

Sohngen, B., and R. Mendelsohn. 2003. An Optimal Control Model of Forest Carbon Sequestration. *American Journal of Agricultural Economics* 85(2):448–57.

Sohngen, B., R. Mendelsohn, and R. Sedjo. 1999. Forest Management, Conservation, and Global Timber Markets. *American Journal of Agricultural Economics* 81(1):1–13.

Wear, D.N., and B.C. Murray. 2004. Federal Timber Restrictions, Interregional Spillovers, and the Impact on U.S. Softwood Markets. *Journal of Environmental Economics and Management* 47:307–30.

CHAPTER 5: IDENTIFYING THE BASELINE CANDIDATES

Murtishaw, S., J. Sathaye, and M. Lefranc. 2005. "Spatial Boundaries and Temporal Periods for Setting Greenhouse Gas Performance Standards." Energy Policy.

CHAPTER 8: APPLYING A LAND-USE OR MANAGEMENT TREND FACTOR

Faeth, P., C. Cort, and R. Livernash. 1994. *Evaluating the Carbon Sequestration Benefits of Forestry Projects in Developing Countries.* Washington, D.C.: World Resources Institute.

Hall, M.H.P., C.A.S. Hall, and M.R. Taylor. 2000. Geographical Modeling: The Synthesis of GIS and Simulation Modeling. *In Quantifying Sustainable Development: The Future of Tropical Economies,* edited by C.A.S. Hall. San Diego: Academic Press.

Hall, C.A.S., H. Tian, Y. Qi, G. Pointius, J. Cornell, and J. Uhlig. 1995. Modeling Spatial and Temporal Patterns of Tropical Land-Use Change. *Journal of Biogeography* 22:753–57.

Scotti, R. 2000. Demographic and Ecological Factors in FAO Tropical Deforestation Modeling. *In World Forests from Deforestation to Transition?* edited by M. Palo and H. Vanhanen. Dordrecht: Kluwer Academic Publishing.

United Nations Food and Agriculture Organization (UN FAO). 1993. Forest Resources Assessment 1990—Tropical Countries. *Forestry Papers* 112, Rome.

CHAPTER 9:   CARBON STOCK QUANTIFICATION

Brown, S. 1999. Land-Use and Forestry Carbon-Offset Projects. Winrock International paper prepared for US AID Environmental Officers Training Workshop, September.

Brown, S., O. Masera, and J. Sathaye. 2000. Project-Based Activities. *In Land Use, Land-Use Change, and Forestry; Special Report to the Intergovernmental Panel on Climate Change,* edited by R. Watson, I. Noble, and D. Verardo. Cambridge: Cambridge University Press.

Intergovernmental Panel on Climate Change (IPCC). 1997. *Revised 1996 IPCC Guidelines for National Greenhouse Gas Inventories.* Kanagawa: Technical Support Unit, IPCC National Greenhouse Gas Inventories Program.

Intergovernmental Panel on Climate Change (IPCC). 2001. *Good Practice Guidance and Uncertainty Management in National Greenhouse Gas Inventories,* edited by J. Penman et al. Hayama: IPCC National Greenhouse Gas Inventories Programme, Institute for Global Environmental Strategies.

Intergovernmental Panel on Climate Change (IPCC). 2003. *Good Practice Guidance for Land Use, Land-Use Change and Forestry,* edited by J. Penman et al. Hayama: Institute for Global Environmental Strategies (also available at http://www.ipcc-nggip.iges.or.jp).

MacDicken, K.G. 1997. *A Guide to Monitoring Carbon Storage in Forestry and Agroforestry Projects.* Arlington, Va.: Winrock International (also available at www.winrock.org).

Masera, O. R., J. Garza-Caligaris, M. Kanninen, T. Karjalainen, J. Liski, G.J. Nabuurs, A. Pussinen, B.H.J. de Jong, and G.M.J. Mohren. 2003. Modeling Carbon Sequestration in Afforestation, Agroforestry and Forest Management Projects: The CO_2FIX V.2 Approach. *Ecological Modelling* 164:177–99. (The CO_2FIX model is available at http://www2.efi.fi/projects/casfor/.)

Parton, W.J., D.S. Schimel, C.V. Cole, and D.S. Ojima. 1987. Analysis of Factors Controlling Soil Organic Matter Levels in Great Plains Grasslands. *Soil Science Society of America Journal* 51:1173–79.

Pearson, T., S. Walker, and S. Brown. 2005. *Sourcebook for Land Use, Land-Use Change and Forestry Projects.* Washington, D.C.: BioCarbon Fund and Winrock International.

Thornton, P.E., B.E. Law, H.L. Gholz, K.L. Clark, E. Falge, D.S. Ellsworth, A.H. Goldstein, R.K. Monson, D. Hollinger, M. Falk, J. Chen, and J.P. Sparks. 2002. Modeling and Measuring the Effects of Disturbance History and Climate on Carbon and Water Budgets in Evergreen Needleleaf Forests. *Agricultural and Forest Meteorology* 113:185–222.

World Resource Institute / World Business Council for Sustainable Development (WRI/WBCSD). 2003. Measurement and Estimation Uncertainty for GHG Emissions Calculation Tool. Available at www.ghgprotocol.org.

Additional Resources

Brown, S. 1997. Estimating Biomass and Biomass Change of Tropical Forests: A Primer. UN FAO Forestry Paper 134, Rome.

Brown, S. 2002. Development and Comparison of Approaches for Establishing Baselines for Land-Use Change and Forestry Projects. Paper presented to the USDA Symposium on Natural Resource Management to Offset GHG Emissions, November 19–21, Raleigh, N.C.

Cairns, M.A., S. Brown, E.H. Helmer, and G.A. Baumgardner. 1997. Root Biomass Allocation in the World's Upland Forests. *Oecologia* 111:1–11.

Guo, L.B., and R.M. Gifford. 2002. Soil Carbon Stocks and Land Use Change: A Meta-Analysis. *Global Change Biology* 8:345–60.

Hamburg, S.P. 2000. Simple Rules for Measuring Changes in Ecosystem Carbon in Forestry-Offset Projects. *Mitigation and Adaptation Strategies for Global Change* 5:25–37.

Harmon, M.E., and J. Sexton. 1996. *Guidelines for Measurements of Woody Detritus in Forest Ecosystems.* US LTER Publication no. 20. Seattle: US LTER Network Office, University of Washington.

Intergovernmental Panel on Climate Change (IPCC). 1997. *Revised 1996 IPCC Guidelines for National Greenhouse Gas Inventories,* edited by J.T. Houghton et al. Bracknell: U.K. Meteorological Office, IPCC/OECD/IEA.

Lal, R., J.M. Kimble, R.F. Follett, and B.A. Stewart (Eds.). 2000. Assessment Methods for Soil

Carbon. Boca Raton: Lewis Publishers. 696 p.

Malhi,Y., D.D. Baldocchi, and P.G. Jarvis. 1999. The Carbon Balance of Tropical, Temperate and Boreal Forests. *Plant Cell and Environment* 22:715–40.

Noriega, F.R. 2002. Modeling Deforestation Baselines Using LUCS for the Noel Kempff, Guaraquecaba, and Chiapas Regions. Submitted to Winrock International.

Tipper, R., and B.H. de Jong. 1998. Quantification and Regulation of Carbon Offsets from Forestry: Comparison of Alternative Methodologies, with Special Reference to Chiapas, Mexico. *Commonwealth Forestry Review* 77(3):219–28.

CHAPTER 10: MONITORING AND QUANTIFYING GHG REDUCTIONS

Intergovernmental Panel on Climate Change (IPCC). 2003. *Good Practice Guidance for Land Use, Land-Use Change and Forestry,* edited by J. Penman et al. Hayama: Institute for Global Environmental Strategies (also available at http://www.ipcc-nggip.iges.or.jp).

International Panel on Climate Change (IPCC). 2006. *2006 IPCC Guidelines for National Greenhouse Gas Inventories,* edited by S. Eggleston et al. Hayama: Institute for Global Environmental Strategies.

MacDicken, K.G. 1997. *A Guide to Monitoring Carbon Storage in Forestry and Agroforestry Projects.* Arlington, Va.: Winrock International (also available at www.winrock.org).

Pearson, T., S. Walker, and S. Brown. 2005. *Sourcebook for Land Use, Land-Use Change and Forestry Projects.* Washington, D.C.: BioCarbon Fund and Winrock International.

CHAPTER 11: CARBON REVERSIBILITY MANAGEMENT PLAN

Subak, S. 2003. Replacing Carbon Lost from Forests: An Assessment of Insurance, Reserves, and Expiring Credits. *Climate Policy* 130:1–17.

PART III: EXAMPLE NIPAWIN AFFORESTATION PROJECT

Boehm, M., B. Junkins, R. Desjardins, S. Kulshreshtha, and W. Lindwall. 2004. Sink Potential of Canadian Agricultural Soils. *Climatic Change* 65:297–314.

Campbell, C.A., R.P. Zenter, S. Gameda, B. Blomert, D.D. Wall. 2002. Production of annual crops on the Canadian prairies: Trends during 1976-1998. Can. J. Soil Sci. 82: 45–57.

Environment Canada. 2004. *Canadian Climate Normals 1971–2000: Nipawin, Saskatchewan.* Created 6/21/02, modified and reviewed 2/25/04. Available at http://www.climate.weatheroffice.ec.gc.ca/climate_normals/results_e.html

Graham, P.E. 2006. Offset System Quantification Protocol for Afforestation Projects. Unpublished report. Ottawa: Natural Resources Canada.

Heller, M.C., G.A. Keoleian, and T.A. Volk. 2003. Life Cycle Assessment of a Willow Bioenergy Cropping System. *Biomass and Bioenergy* 25:147–65.

Intergovernmental Panel on Climate Change (IPCC). 2003. *Good Practice Guidance for Land Use, Land-Use Change and Forestry,* edited by J. Penman et al. Hayama: Institute for Global Environmental Strategies (also available at http://www.ipcc-nggip.iges.or.jp).

Joss, B.N., R.J. Hall, D.M. Sidders, and T.J. Keddy. 2005. Site Suitability for High-Yield Hybrid Poplar. Map product. Edmonton: Northern Forest Service, Natural Resources Canada.

Lemprière, T.C., M. Johnston, A. Willcocks, B. Bogdanski, D. Bisson, M. Apps, and O. Bussler. 2002. Saskatchewan Forest Carbon Sequestration Project. *Forestry Chronicle* 78:843–49.

Li, Z., W.A. Kurz, M.J. Apps, and S.J. Beukema. 2003. Belowground Biomass Dynamics in the Carbon Budget Model of the Canadian Forest Sector: Recent Improvements and Implications for the Estimation of NPP and NEP. *Canadian Journal of Forest Research* 33:126–36.

Natural Resources Canada. 2006. National Forest Inventory: National Forest Information System. Last updated 5/29/06. Available at https://nai.nfis.org/nai-home/public/splash.do.

Niu, X., and S.W. Duiker. 2006. Carbon Sequestration Potential by Afforestation of Marginal Agricultural Land in the Midwestern U.S. *Forest Ecology and Management* 223:415–27.

Peterson, E.B., G.M. Bonnor, G.C. Robinson, and N.M. Peterson. 1999. Carbon Sequestration Aspects of an Afforestation Program in Canada's Prairie Provinces. Report submitted to Joint Forest Sector Table/Sinks Table, National Climate Change Process, Nawitka Renewable Resource Consultants, Victoria, B.C.

Prairie Farm Rehabilitation Administration. 2001. Prairie Land and Water Resources: The Last 100 Years. Soil Resources. Regina, Saskatchewan. Accessed 6/26/06. Available at http://collections.ic.gc.ca/soilandwater/lr4.htm.

(S&T)[2] Consultants Inc. 2005. Documentation for Natural Resources Canada's GHGenius Model 3.0. Delta, B.C.

Soil Landscapes of Canada Working Group. 2006. Soil Landscapes of Canada v3.1. Agriculture and Agri-Food Canada. (digital map and database at 1:1 million scale). Available at http://sis.agr.gc.ca/cansis/nsdb/slc/v3.1/intro.html.

93

Statistics Canada. 1996. *Census of Agriculture. Data Tables* (e-mail correspondence with Debbie Jacobs, account executive with the Western Region and Northern Territories, received 7/5/06).

Statistics Canada. 2001. *Census of Agriculture: Data Tables.* Modified 6/28/02. Available at http://www.statcan.ca/english/freepub/95F0301XIE/tables.htm

Statistics Canada. 2003. *The Daily: 2001 Census of Agriculture—Canadian Farm Operations in the 21st Century.* Modified 1/9/03. Available at http://www.statcan.ca/Daily/English/020515/d020515a.htm

U.S. Environmental Protection Agency (US EPA). 2006. *Non-CO$_2$ Gases and Carbon Sequestration: Conversion Units.* Updated 3/7/06. Available at http://www.epa.gov/nonco2/units.html.

Zentner, R.P., D.D. Wall, C.N. Nagy, E.G. Smith, D.L. Young, P.R. Miller, C.A. Campbell, B.G. McConkey, S.A. Brandt, G.P. Lafond, A.M. Johnston, and D.A. Derksen. 2002. Economics of Crop Diversification and Soil Tillage Opportunities in the Canadian Prairies. Agronomy Journal 94:216–230.

GLOSSARY

United Nations Food and Agriculture Organization (UN FAO). 2000. *Global Forest Resources Assessment 2000.* Rome: UN FAO.

Intergovernmental Panel on Climate Change (IPCC). 2003. *Good Practice Guidance for Land Use, Land-Use Change and Forestry,* edited by J. Penman et al. Hayama: Institute for Global Environmental Strategies (also available at http://www.ipcc-nggip.iges.or.jp).

ANNEX A: LIFE-CYCLE ASSESSMENTS AND UPSTREAM AND DOWNSTREAM EFFECTS

Heller, M., G. Keoleian, and T. Volk. 2003. Life Cycle Assessment of a Willow Bioenergy Cropping System. *Biomass and Bioenergy* 25:147–65.

Intergovernmental Panel on Climate Change (IPCC). 1997. Revised 1996 *IPCC Guidelines for National Greenhouse Gas Inventories,* edited by J.T. Houghton et al. Bracknell: U.K. Meteorological Office, IPCC/OECD/IEA.

Other Contributors

Many of the contributors in this list helped us think through the questions and concerns around the LULUCF accounting issues while we were still working on the generic Project Protocol. Much of the information collected then has also passed into this document and we would like to thank them for their contributions also. Contributors are listed with their affiliations at the time of their contribution.

Steve Apfelbaum, Applied Ecological Services

Cheryl Miller, Applied Ecological Services

John O. Niles, Climate, Community & Biodiversity Alliance

Sonal Pandya, Conservation International

Michael Totten, Conservation International

Ben de Jong, El Colegio de la Frontera Sur

Richard Tipper, Edinburgh Centre for Carbon Management (ECCM)

George Fowkes, Future Forests

Paul Norrish, Future Forests

David Brand, Hancock Natural Resource Group

Promote Kant, Indian Council of Forestry Research and Education

Jayant Sathaye, Lawrence Berkeley National Laboratory

Jeff Fielder, Natural Resources Defense Council

Jane Ellis, Organization for Economic Co-operation and Development

Laurie Wayburn, Pacific Forest Trust

Brian Murray, Research Triangle Institute (RTI)

Allan Sommer, RTI

Neil Sampson, Sampson Group

Erik Firstenberg, The Nature Conservancy (TNC)

Patrick Gonzalez, TNC

Ellen Hawes, TNC

Zoe Kant, TNC

Tia Nelson, TNC

Bill Stanley, TNC

Michelle Manion, Union of Concerned Scientists

Kate Bickley, US Environmental Protection Agency (US EPA)

Ben DeAngelo, US EPA

Maurice Le Franc, US EPA

Sandra Brown, Winrock International

Ian Nobel, World Bank

Ordering publications

WRI

Hopkins Fulfillment Service

Tel: (1 410) 516 6956

Fax: (1 410) 516 6998

e-mail: hfscustserv@mail.press.jhu.edu

Publications can be ordered from WRI's secure online store:

http://www.wristore.com

Disclaimer

This document, designed to promote best practice GHG project accounting and reporting, has been developed through a globally diverse multi-stakeholder consultative process involving representatives from business, nongovernmental organizations, government, academics, and other backgrounds. While WRI encourages the use of the LULUCF Guidance for GHG Project Accounting, its application and the preparation and publication of reports based on it are the full responsibility of its users. In particular, use of the LULUCF Guidance does not guarantee a particular result with respect to quantified GHG reductions, or acceptance or recognition of quantified GHG reductions by GHG programs. Neither WRI, nor the individuals who contributed to the LULUCF Guidance assume responsibility for any consequences or damages resulting directly or indirectly from its use and application.

Copyright ©World Resources Institute

October 2006

ISBN 1-56973-631-6

Library of Congress Control Number: 2006936591

Printed in USA

FSC
Mixed Sources
Product group from well-managed forests, controlled sources and recycled wood or fiber
Cert no. SW-COC-1530
www.fsc.org
© 1996 Forest Stewardship Council

Printed with linseed-oil-based inks on, elemental chlorine free paper Chorus Art Silk containing 50% recycled content including 25% post consumer content. Using this paper created the following benefits: 5.85 trees not cut down; 16.89 lbs. waterborne waste not created; 2,485 gallons wastewater flow saved, 275 lbs. solid waste not generated, 541 lbs. net greenhouse gases prevented and 4,144,175 BTUs energy not consumed.